PHILOSOPHY OF MEDICINE

PHILOSOPHY OF MEDICINE

Alex Broadbent

OXFORD
UNIVERSITY PRESS

OXFORD
UNIVERSITY PRESS

Oxford University Press is a department of the University of Oxford. It furthers
the University's objective of excellence in research, scholarship, and education
by publishing worldwide. Oxford is a registered trade mark of Oxford University
Press in the UK and certain other countries.

Published in the United States of America by Oxford University Press
198 Madison Avenue, New York, NY 10016, United States of America.

© Oxford University Press 2019

Library of Congress Cataloging-in-Publication Data
Names: Broadbent, Alex, 1980- author.
Title: Philosophy of medicine / Alex Broadbent.
Description: New York : Oxford University Press, 2018. |
Includes bibliographical references.
Identifiers: LCCN 2018018337 (print) | LCCN 2018020331 (ebook) |
ISBN 9780190612177 (online content) | ISBN 9780190612153 (updf) |
ISBN 9780190612160 (epub) | ISBN 9780190612146 (pbk. : alk. paper) |
ISBN 9780190612139 (cloth : alk. paper)
Subjects: LCSH: Medicine—Philosophy. | Medical ethics.
Classification: LCC R723 (ebook) |
LCC R723 .B67 2018 (print) | DDC 610.1—dc23
LC record available at https://lccn.loc.gov/2018018337

1 3 5 7 9 8 6 4 2

Paperback printed by Webcom, Inc., Canada
Hardback printed by Bridgeport National Bindery, Inc., United States of America

For Theodore

CONTENTS

PART B
WHAT SHOULD WE THINK OF MEDICINE?

ACKNOWLEDGMENTS

I wrote the first draft of this book in serial form, committing myself to generating a chapter every week or two in the first part of 2017 by promising to circulate the results among interested readers. I did not quite keep the pace up but nonetheless had a draft by April of the same year, as well as a rich collection of comments. I am thus very grateful to these readers, for their comments and their imagined anticipation, on my part; convincing myself that they were eagerly awaiting the next chapter provided motivation, and I also find it easier to write with a concrete audience in mind, rather than an abstract one. Readers included Richard Broadbent, Thomas Cunningham, Adrian Erasmus, Arantza Etxeberria, Olaf Dammann, Halley Faust, Katherine Furman, Daniel Goldberg, Sander Greenland, Bennett Holman, Dieneke Hubbeling, John Huss, Ian Kidd, Robert Kowalenko, James Krueger, Jonathan Livingstone-Banks, Hanna van Loo, Thomas Maloney, Mat Mercuri, Albert Mosley, Dominik Nowak, Nancy Nyquist Potter, Rose Richards, Thomas Schramme, Reinier Schuur, Jonathan Sholl, Jeremy Simon, Alok Srivastava, Jacob Stegenga, Zia Taylor, Sean Valles, and Sander Werkhoven. I'm immensely grateful to all of them for volunteering their time to read, and especially those who also offered comments. More extended conversations occurred with Jonathan Fuller, Benjamin Smart, Jacob Stegenga, and Jan Vandenbroucke, and these were very important for the resulting book. My thanks also go to Likhwa Ncube, who, as my research assistant, conducted research that was essential to the success of this project, and who helped me through our discussions with many of the book's ideas. I apologize if I have missed anyone from these lists of readers and discussants; there have been many and I have tried to keep records, but recordkeeping is not my forte. Thea de Wet introduced me to Roy Porter's work, and lent me her copy of *The Greatest Benefit to Mankind*, which I still intend to return. I am particularly indebted to Chadwin Harris and Thaddeus Metz, colleagues at the University of Johannesburg who delivered critical responses to my inaugural lecture, and then developed these responses in print for a special section of the *Journal of Medicine and Philosophy*. I am further grateful to the various audiences that allowed me to present versions of ideas in this book, including the Philosophical Society of Southern Africa, Bergen University Philosophy Department, Cambridge Department of History and Philosophy of Science, the Oxford Philosophy of Medicine Research Group, the International Philosophy of Medicine Roundtable, and the CauseHealth project. I am grateful to the Johannesburg Institute for Advanced Studies for providing me with a congenial venue for a writing retreat during the month of July, when the manuscript was finalized. I am further grateful to Tshilidzi Marwala and Angina Parekh, my superiors (institutionally speaking), who gave me permission to back off from my duties as a dean during the crucial month, as well as to Maria Frahm-Arp and Kammila Naidoo, my vice-deans, for covering for me so effectively that I got away with checking my university email account only three times during the entire month. Lorna Singh deserves special mention for support in numerous respects, especially managing and protecting my time, ensuring a congenial working environment, and pretending so convincingly that I am in an important meeting when I am having a nap. Finally I thank my family, James, Sylvia, Sheila, and Gugu, and especially my wife, Nicole, for such a warm and supportive home life, and thus for nourishing my heart in a way that the professional side of life never can.

Chapters 2 and 3 draw upon material previously published in the *Journal of Medicine and Philosophy* (Broadbent 2018a; Broadbent 2018b). Chapter 4 draws upon material previously published in *The British Journal for the Philosophy of Science* (Broadbent forthcoming). Both journals are published by Oxford University Press and I am grateful for permission to partially reproduce and develop them here. Chapter 9 includes ideas and material developed in an article in *The Conversation Africa* called "It will take critical, thorough scrutiny to truly decolonise knowledge," which can be found here: https://theconversation.com/it-will-take-critical-thorough-scrutiny-to-truly-decolonise-knowledge-78477. I am grateful to *The Conversation* for allowing me to partially reproduce and develop the material here.

INTRODUCTION

I.1 TWO BIG QUESTIONS

This book sets out to answer two big questions about medicine: "What is medicine?" and "What attitude should we adopt toward it?" The answer to the first question informs the answer to the second. Through answering these two questions, the book provides a unifying vision of the field that is at once useful for someone looking for a way into the field, and novel enough to interest the expert. Although much of what has been written by contemporary philosophers of medicine speaks to one or both questions, neither question has received a book-length philosophical treatment.

Yet these questions are more pressing than they have ever been, given the proliferation of medical paradoxes. Medicine cures more, and people are healthier; yet medicine is criticized more, and people trust it less. One tradition attains a position of global dominance that is historically unprecedented, and enjoys a unique reputation for effectiveness; yet other traditions become *more* popular, not less, even in the regions of the developed world where Mainstream Medicine (as I call it) originates. Above all, medicine has existed for millennia without being very good at curing people, and it remains at best patchy in this regard; yet the medical profession is typically held in high esteem. What is this thing, and what are we to make of it—not just in the abstract, but when we are confronted with an ailment and need to decide what to do about it, or with a piece of medical advice that we are uncertain about following? These are philosophical questions, but they are not *merely* philosophical: they are personal, political, and practical, too.

In the introduction to his magnum opus, the great historian of medicine, Roy Porter, writes:

> I say here, and I will say many times again, that the prominence of medicine has lain only in small measure in its ability to make the sick well. This was always true, and remains so today.
>
> (Porter 1997, 6)

The subsequent 800-plus pages emphasize the fact that, for most of history, medicine has possessed remarkably few cures. Yet the point is not to ridicule medicine. The point is to see it clearly, without either Whiggish triumphalism or sardonic sneer, and so to understand it: as it was, and as it is today.

My own book is also an attempt to understand medicine, and it is also in part a reaction to the commonness with which medicine fails to heal the sick. Yet it is not a skeptical or cynical treatment of medicine, any more than Porter's history. Rather, it is an attempt to understand medicine, as it is rather than as we wish it were, and to provide a philosophical response to some of the same questions about medicine that Porter's history illuminates.

In particular, two questions stand out. The first is, *what is medicine?* Since settled communities emerged 10,000 to 12,000 years ago, there have been individuals or professional groups recognized as possessing expertise relating to sickness and health. Sometimes the medical role has been mixed with that of the priest (as in ancient Egypt), sometimes with that of the philosopher (as recommended by Galen, who shaped 1,000 years of Western practice), and, in the contemporary world, with that of the empirical scientist. This multitude of traditions has no obvious concepts or practices in common. Conceptually, some are naturalistic, some supernatural; some holistic, some not; some explain illness in terms of factors that are external to the physical body, some explain illness internally; some see health and illness as personal and unique, others emphasize disease entities that are common to many individuals; and many traditions incorporate elements of both sides of each of these dichotomies, either harmoniously or discordantly. Practically, the recommendations of various traditions are even more extravagantly various, and for almost every practice that has been recommended as promoting health or healing illness in a given circumstance, another tradition or practitioner has prohibited it or recommended something incompatible in the same circumstance. What ties this cacophony together, and enables us to recognize it all as medicine, of one sort or another—if, indeed, anything does?

My answer is that the goal of medicine is to heal the sick, but the core business of medicine is to understand and predict health and disease. I call this view the *Inquiry Thesis,* because it says that medicine is fundamentally an inquiry with a purpose. The contrast is the *Curative Thesis,* which says that medicine is simply the sustained and organized effort to heal the sick, by whatever means. The two may not look entirely dissimilar, and indeed they agree on the goal of medicine, namely curing the sick; but they disagree on its nature, and they have quite different consequences for the way we understand medical successes and failures. Much of the book is devoted to arguing that only the Inquiry Thesis, and not the Curative Thesis, can reconcile the miserable track record of our efforts to cure, in most times and places, with the generally high status accorded to the medical profession.

The second question that stands out is, *what attitude should we adopt toward medicine?* The frequent inability of medicine to heal the sick is obviously relevant here too. Even pointing it out can sound like an attack on medicine, an attempt to undermine confidence in it.

You might take courage from the fact that not all medicine is the same, and argue that some medical tradition, or some disciplines within some traditions, are in fact successful. You might then argue that we should be confident and trusting insofar as we have reason to think that a given tradition or discipline or intervention is successful. This is *Medical Whiggishness*, the view, roughly, that the present successes contrast with past failures, and that we are on the royal road of progress. Or you might concede that the abysmal track record of curative medicine admits of no systematic exceptions, and thus that we should place low confidence in medicine quite generally. This, roughly, is *Medical Nihilism*. I am personally tempted by both reactions, and I explore both in this book. But I reject them both. The former, optimistic, reaction forces me to decide what counts as good evidence, and I don't think this can be done, at least not in a way that suits this purpose. And I reject the latter, pessimistic, view because it condemns too much of the actual reliance that I place on medicine as irrational, for me to comfortably accept it. I suspect that my reactions are typical, and I argue that they are reasonable.

Being unwilling to choose between Medical Whiggishness and Medical Nihilism, I propose to endorse an attitude that I call *Medical Cosmopolitanism*. It is an implementation of Kwame Anthony Appiah's general proposal of Cosmopolitanism as a basis for civilized society, and particularly as an approach to ethical disagreements, in a world of great difference of opinion. I see the Cosmopolitan approach as helping to make sense of both medical disagreements and, more broadly, the persistence of medicine in the face of its curative ineffectiveness. I thus take Cosmopolitanism as a starting point, which informs my approach throughout the book, and not only its conclusion; but I will also provide arguments for it. This approach is not circular, since the arguments could be provided even if my attitude (or at least, the attitude to which I aspire) were dogmatic and epistemically hubristic rather than inquisitive and epistemically humble.

I.2 LIMITATIONS

There are two important philosophical aspects of medicine that this book does not cover. The first is the philosophy of psychiatry. Some reference will be made to mental illness, so the omission is not complete. Moreover, the guiding question "What is medicine?" no doubt would benefit from discussion of topics such as madness, mental health and illness, psychology, psychiatry, psychoanalysis, and so forth. However, the philosophy of psychiatry and its literature are large enough in their own right to merit separate treatment; indeed, there is more extant work on the philosophy of psychiatry than on the philosophy of medicine per se. For these reasons, I have decided not to dedicate any whole chapters to topics falling under the general area of mental health and medicine.

The second limitation of the book, already mentioned, is its omission of normative ethical considerations: questions concerning whether physicians may assist patients wishing to commit suicide, for example, or how conflicts between patient and physician values ought to be resolved. This is partly because my own interests are in metaphysical and epistemological questions: questions about what there is and how we know about it. But it is mainly because normative ethical questions relating to medicine have attracted very substantial attention for a very long time (Hippocrates was so concerned about ethical questions that he came up with an Oath), and since the rise of contemporary Mainstream Medicine they have been the subject of many books. This means, once again, that a separate treatment would be merited. Trying to squeeze five decades of modern bioethics into a chapter or two would serve little purpose. And I am not willing to let it take up half the book: there are other fundamental topics in the philosophy of medicine, concerning its nature and our attitude toward it, and they need room to grow.

This is not to say that normative considerations will not make appearances: it is impossible to keep them out, given that medical activity usually has a purpose, and that medical traditions themselves often seek to prescribe some purpose or purposes, and proscribe others. But my concern will not be in normative biomedical ethics: that is, the important process of thinking through, whether in a principled or casuistic fashion, what courses of action are mandatory, permissible, or impermissible under given circumstances. Likewise I will not be discussing those well-known metaphysical questions that derive a large part of their interest from ethical considerations, such as at what stage of development a fetus becomes a person with a right to life, or whether there is such a thing as free will to ground notions like autonomy and informed consent. These topics receive very substantial treatments elsewhere, and moreover they do not much help me make progress with the two big questions I am hoping to answer.

PART A

WHAT IS MEDICINE?

Chapter 1

Varieties of Medicine

1.1 EARLY MEDICINE

The point of this book is not to offer a history of medicine. Nonetheless, I believe that some historical awareness of at least some medicine is important for philosophy of medicine in general. It is certainly important for the next three chapters. So in this section, I summarize extremely crudely and briefly some of the main elements of Roy Porter's substantial medical history of humanity (Porter 1997), especially relying on Chapters 1 and 2, which deal with the beginnings of medicine; and for later periods (where the magnum opus becomes very detailed) on his shorter summary, published in the year of his death (Porter 2002). Both the short and the long books are highly recommended. I will not litter the following text with attributions and references, but I am not ashamed to emphasize, once again, the heavy reliance that I place on Porter's work in this section especially, and more generally in my historical understanding of medicine.

Our ancestor *Australopithecus* a kind of ground-dwelling ape, emerged in southern Africa about 5 million years ago. A more recent ancestor, *Homo erectus*, emerged about 2 million years ago, as a carnivorous hunter, and spread from Africa to Asia and Europe. Sometime after that, about 1 million years ago, *Homo sapiens*—modern humans—emerged.

Early humans were nomadic hunter-gatherers. They had tough lives, and they were vulnerable to injury, starvation, the weather, and predators. However, early humans did not suffer much from infectious disease of the kind we are so familiar with. They might suffer diseases caught by eating diseased animals, such as anthrax, or picked up from the soil, or through infected wounds. But they did not suffer the scourges that have since afflicted settled societies, nor indeed those afflictions, major and minor, that continue to afflict us today. Cavemen did not get colds.

This is because infectious diseases require large pools of hosts living in close proximity. As hunter-gatherers forming groups of 30 to 40, humans did not form large enough pools. Nor did they live in close proximity to species that did. That changed as the project of global colonization began to reach its limit, and virgin land for hunting and gathering was no longer easily had. This happened around the end of the last Ice Age, about 12,000 to 15,000 years ago.

It is not clear that this amounted to progress. In the words of Roy Porter:

> Contrary to the Victorian assumption that farming arose out of mankind's inherent progressiveness, it is now believed that tilling the soil began because population pressure and the depletion of game left no alternative: it was produce more or perish.
>
> (Porter 1997, 17)

Whatever the reason, many hunter-gatherers became farmers. There remain, to this day, some exceptions: the San people of the Kalahari, the Aboriginals of Australia, some Native Americans. But in most of the world, including in Africa, nomadic existences gave way to agricultural ones.

Agriculture enables the production of more food, enabling an increase in the quantity of human life, but it may also lower quality of life. Variety of food often suffers, and with it, nutrition. Dependence on nutrient-lacking cereal crops, such as maize or rice, has sometimes led to deficiencies such as pellagra. It was reliance on maize that led to Richard Goldberger's discovery that pellagra was a disease of nutrition (Morabia 2004, 48–52). Agriculture also requires intense manual labor, and favors dwelling in close physical proximity to other people, in a fixed location. These create the fertile fields for infectious disease to flourish.

As well as creating a fertile ground for infectious disease, agriculture supplies the seed: intimate contact with herd animals, from whom infectious diseases can make a jump across species. It is from animals that our infectious diseases mostly come. We share over 65 diseases with dogs: measles is derived from canine distemper. Cattle gave us tuberculosis, horses gave us the common cold (rhinovirus), poultry and pigs gave us influenza. These jumps continue to happen periodically: think of HIV, so-called bird flu, and Ebola. Agriculture also increased the quantity of suitable physical environments for insect hosts, such as swamps for mosquitoes. Malaria, the single most damaging disease in human history, came with settlement, as forests were cleared, irrigation was developed, and swamps were created. These developments also created new environments for worms and other parasites.

Disease was thus an affliction of civilization, and in this, genesis stories of fall from grace have some resonance with the truth. Agriculture enabled more people to live, but they were sicker and less well nourished, as evidenced by the fact that Paleolithic skeletons are generally somewhat taller than their Neolithic descendants.

Hunter-gatherers are unlikely to have had medical experts, instead dealing with their sick collectively, since illness would have been very much a collective concern for groups whose mobility was essential to their survival. Medical anthropologists identify two "sick roles": that of the childlike dependent, who is nurtured and cared for until health returns or death comes; and that of the outcast, who is ejected and left behind. Hunter-gatherers would probably have abandoned their sick more commonly than not. But stationary communities with larger numbers of people falling ill were more likely to care for their ill members. Shamans, diviners, and medicine men arose; it is not clear exactly when.

Most medical encounters, and most medical practices and practitioners, are a mystery to us. We know "literally nothing" about Hippocrates, "perhaps the most celebrated physician ever" (Porter 1997, 13). What we know of very early medical practices is pieced together from archeological remains and cave paintings, and also from studying the few remaining nomadic peoples of the world. But around 2000 B.C.E., medicine began to be recorded in writing. This not only affects our historical knowledge of medicine, but also affected medicine itself. "Learned medicine" arose, with the ability to transmit, exchange, and accumulate medical learning across distance and time using written records; and the scope for hierarchy and social organization of medicine correspondingly increased.

There have been several learned medical traditions, and, besides Mainstream Medicine, several persist in some form, notably Indian and Chinese medicines. Both these systems tended to glorify tradition, as did the Hippocratic/Galenic school in the West. Neither laid emphasis on innovation (although developments did in fact occur) and, in due course, both experienced tense encounters with Western "scientific medicine," eventually forcing some change upon them (Porter 1997, 135). We will discuss some of these further in section 1.4.

1.2 MAINSTREAM MEDICINE

The far origins of what I call "Mainstream Medicine" are in ancient Egypt, which influenced Ancient Greek medicine. However, Greek medicine did not resemble Egyptian; it is reasonably thought of as original, and thus the more appropriate point to pick as the origin of our contemporary dominant system. Ancient Greek medicine was distinctive for its naturalism and characteristically Greek intellectual openness.

Hippocrates, perhaps the most famous doctor ever, is said to have been born on the island of Cos around 460 B.C.E. Although there are about 60 works attributed to him, they are clearly written by a number of different authors, and his authorship is similar to that of Homer over the *Iliad*. Hippocrates espoused a humoral theory of medicine, in which health arose from balance in the four humors, or fluids, within the body: black bile, phlegm, choler (yellow bile), and blood. These four fluids paralleled the four natural elements identified by Greek science (earth—black bile, water—phlegm, fire—choler, air—blood).

According to Porter, "The appeal of the humoralism which dominated classical medicine and formed its heritage lay in its comprehensive explanatory scheme, which drew upon bold archetypal contrasts (hot/cold, wet/dry, etc.) and embraced the natural and the human, the physical and the mental, the healthy and the pathological" (Porter 2002, 29). It is important to see that while the Greeks were naturalistic, they nonetheless explained health and illness in what we might call cosmic terms, in common with almost all other medical traditions: that is, by relating health and illness to the wider world, and humankind's place in it. Hippocrates also famously urged consideration of the environment and lifestyle in understanding the health of an individual. Some classic epidemiology textbooks also credit Hippocrates with being the forerunner of epidemiology (e.g., MacMahon and Pugh 1960), although the epidemiologist–historian Alfredo Morabia argues that Hippocrates' interest was in the effect of the environment on individuals, for clinical purposes, and thus that there is not enough evidence of population thinking to support this attribution (Morabia 2004, 93).

The Hippocratic mindset influenced medicine right up to the early 20th century, although there were sometimes rivals. Besides the naturalism of the Hippocratic tradition, and its humoral theory of disease, an even longer-lasting legacy for future centuries was the fact that it contained an explicit ethical code, the Hippocratic Oath. The Oath established an ideal of medical practice that was explicitly principled, and the contents were selfless, patient-centered, and non-interventionist. The principle *primum non nocere* (first, do no harm) distinguished upstanding medical men, concerned with doing right by their patients and upholding the standards of their profession, from greedy quacks promising miracle cures.

It is obvious from this description that contemporary Mainstream Medicine is in some respects a break from the Hippocratic tradition. Wonder cures have become prized (maybe because they have finally become available), and discovering them is lucrative; and surgery, which always involves a first harm (a cut) and was anathema to Hippocrates (who advised those interested in surgery to go to war), now stands at the pinnacle of the medical profession. There have been a number calls to return to a more Hippocratic ideal (whether or not it is given that name), for example in the bioethical codes made popular by Beauchamp and Childress (2013), and more recently by Jacob Stegenga, whose call for "gentle medicine" (Stegenga 2018) we will consider in Chapter 5.

Hippocrates' influence was not direct: the major direct influence on medieval medicine was the Roman physician–philosopher, Galen. In contrast to the "shadowy" Hippocrates, Galen's life is well documented and his own voluminous writings survive. Also contrasting to Hippocrates, Galen was an egotist, predicting his own influence on future medicine and declaring himself to have set medicine on the right path, to which Hippocrates had pointed the way.

Galen emphasized that the medical man should be a man of philosophy, armed not merely with practical skills but with logic, physics, and ethics. Galen performed animal dissections, human dissections being unacceptable, and worked on the assumption that humans and animals were anatomically identical—an assumption leading to errors.

Galen's work dominated European medicine for the next millennium, partly no doubt because there was no significant advance during that period. During the Dark Ages, there were no learned men left in Europe outside monasteries and the Church. Medicine became their preserve. The more advanced Islamic world was key in sustaining the Galenic corpus: scholars in what are now Syria, Iraq, Iran, Egypt, and Spain translated and developed the work of Galen.

In the 12th century, when Scholasticism was new and the first European universities were founded, professional medicine in Europe began to recover, starting with the retranslation of texts from Islamic sources at Salerno (Italy). Medical education took seven years of learning based on set texts and debate, and required broad philosophical knowledge. It was this that distinguished the learned physician from the quack. There were not many men (women were not permitted to train) who lived up to the Galenic ideal, with apprenticeship and experience remaining central in the actual education of most practicing physicians; but the ideal was there nonetheless.

At this point, the story becomes multithreaded. As I have already suggested, the Hippocratic ideal remained, in many ways, dominant through the 19th and even into the 20th century. However, the various strands that make up the medical profession were not and are not all encompassed by this ideal. Medicine has sought to arrange itself in a hierarchy, with physician on top; and this hierarchy has been repeatedly challenged. Sometimes, physicians have adopted the recommendations of the challengers; at other times, they have not.

From the early Renaissance, anatomical knowledge started to be regarded as essential. The first *recorded* public human dissection (there is reputed to have been a much earlier one in Alexandria) was

by Mondino de' Luzzi in 1315. Public dissection, as spectacle and education combined, became widespread over the next two centuries.
Initially the dissection was a show by the teacher, who, through a performance that the student might not even be able to see very well,
illustrated the correctness of Galenic teachings.

The next important figure is Andreas Vesalius (1514–64), who
pressed the importance of learning through direct observation. He
challenged Galenic teaching, in particular his assumption that animal
and human anatomy are identical. Vesalius suddenly made it acceptable and desirable to achieve new anatomical findings, and this led to
an explosion of anatomical knowledge.

One consequence of greater anatomical knowledge was a gradual
replacement of the humoral theory. However, it must be stressed
that anatomical knowledge did not necessarily equate to knowledge
of the function that the body part in question performed. Knowing
that an organ exists, and where it is, and being able to describe it,
are most definitely not the same as knowing what it does. Gabriele
Falloppio, after whom the tubes connecting ovaries to womb are
named, published details of these tubes along with various other
observations of skull, ear, and female genitals in 1561; but the function of the fallopian tubes was only understood two centuries later.

One of the most significant pieces of anatomical understanding
arising out of the new anatomical science was William Harvey's demonstration of the circulation of blood. He was convinced of some sort
of circulation as early as 1603, but published his full findings in 1628.
Galen had held that blood was produced in the liver and carried in
veins to the rest of the body, which consumed it; and that the heart
pumped air through the arteries. Galen held, too, that blood passed
from the right ventricle to the left, something that Colombo of Spain
had already empirically demonstrated was false. Harvey demonstrated
that blood circulated with several clever arguments and experiments,
even though he could not see capillaries, and thus could not entirely

complete the picture. One argument was that the volume of blood pumped out of the heart each day was in the hundreds of gallons, far more than could possibly be produced or consumed elsewhere in the body. Another was that the veins in a forearm did not swell if it was so tightly bound as to cut off both veins and arteries, but did swell if the ligature were loosened so that the veins but not the arteries were blocked.

These inquiries look thoroughly scientific, and indeed it is fair to call them that. However, it is also important to realize that Harvey remained inspired by larger philosophical ideas. The idea of circulation reflected a commitment to the Aristotelian idea of circular perfection.

Perhaps this does not make them unscientific. A commitment to a spiritual or quasi-spiritual (that is, overarching, super-empirical) ideal is common to many of the great medical scientists, as it was to Newton, Descartes, and many other Enlightenment thinkers. In my mind it is one of the most interesting and difficult features of medical science, and perhaps of other sciences too (physicist–mystics seem especially common). The most technical or empirical thinkers are often mystical about the fundamentals, drawing a sharp line between those things that are to be scrutinized with extreme dispassion and rationality and those that cannot be asked about by science—that is, cannot be scrutinized in this way. They must be handled by philosophy or religion. I have seen this attitude in contemporary medical scientists too (VanderWeele 2015, 448–458), and it raises obvious questions—in particular, how one is to know whether an apparently unanswerable question is one that should be approached by other means, or simply a normal one, which you just happen to be intellectually or technologically incapable of answering in a rational way.

In the 17th and 18th centuries, medical science took a mechanistic turn, with the influence of Boyle, Hooke, and Descartes. The body was seen as a machine, a system of pipes, pulleys, cogs, and so forth. Again, this idea was combined with mysticism about the fundamentals; just as Newton held that the cause of gravity was not susceptible to inquiry, so the source of animation was ascribed to a soul, not susceptible to empirical investigation. Considerable anatomical knowledge was amassed, especially using the microscope, and the hope emerged of a full understanding of the body–machine. However, experimentation also produced decisive demonstrations that the body is not a machine: lobsters' claws regrow when cut off; divided polyps become two new polyps. By the end of the 18th century, mechanical ideas about the body had given way to vitalistic ideas. The term "biology" was coined around 1800.

Hospitals permitted observations; laboratories permitted experiments, and although they had been around since Boyle and Hooke's time, the laboratory sciences of organic chemistry, microscopy, physiology, and others only took off in the 19th century. Laboratory medical science produced dramatic new knowledge, both of anatomy and of disease. In the latter part of the century, Louis Pasteur showed that certain specific bacteria caused certain specific diseases, notably anthrax and rabies. Robert Koch famously discovered the microorganisms responsible for tuberculosis and cholera in the mid-19th century, and revolutionized medical scientific methodology with his four "postulates," satisfaction of which was necessary and sufficient (he held) to prove that a given microorganism causes a given disease. Koch's postulates remain in use today, somewhat modified for viral agents.

What effect did all of this medical knowledge have on medical practice? Very little. There were some "specifics." The use of cowpox to inoculate against smallpox was driven by William Jenner at the end of the 18th century. But this practice arose from taking seriously the folk observation that milkmaids appeared unusually resistant to smallpox, and not directly from laboratory inquiry or the growth of anatomical knowledge. By the end of the 19th century, Pasteur had developed vaccinations against anthrax and rabies. Anesthesia, through gas, was also developed in the 19th century. But the official pharmacopeia remained largely either inert or harmful, a collection of herbal, mineral, or other remedies, including some that were based on nothing more than ancient superstition (e.g., the use of a stone-like substance found in the stomach of ruminants as an antidote to poison). Surgery remained barbaric: painful and dangerous, given the lack of proper control of infection.

This state of affairs led to medical nihilism: a fatalistic view that medical science could uncover the secret workings of the body, and that the medical man (still a man) could thus understand what was happening to his patient, but that he was essentially powerless to do anything about it. "Scientific medicine" took hold in America in particular, involving the physical examination of the patient and of all her secretions (satirists maintained that the medical profession played in particular on the worries of well-to-do women). The stethoscope was invented in the early 19th century and became the hallmark of the scientific doctor. Not everyone liked scientific medicine, by any means; bedside manner, astute cross-questioning, a capacious memory, and modesty about the cures prescribed remained very attractive in a private doctor well into the 20th century. In the Hippocratic-inspired tradition, a remedy was not expected to work wonders, but rather to support healing, along with various advice about diet and lifestyle.

The 20th century saw the therapeutic revolution that had been hoped for and despaired of for so long. It is important to understand that the "medical revolution" of the 20th century did not consist of developments in one single area of medical science or practice. Chemotherapy was initially seen as highly promising after the trial-and-error discovery of the first true cure of syphilis in 1907, called Salvarsan. (Previously syphilis was treated by applying mercury to sores, which is the epitome of a cure that offers no assistance and brings its own harms.) Early promise disappointed, however, until

1935, when the director of research at the chemical company Bayer discovered that a certain red dye could be used to cure streptococcal infection. Similar drugs were developed for a few other diseases, but chemical "magic bullets" were fewer than had been hoped.

Penicillin was discovered by accident in 1928 but initially not seen as significant. Though not the very first test, the first widely known success occurred treating soldiers suffering from gonorrhea in Africa in 1943. Many new drugs were discovered in the 1950s and 1960s, including vaccines that were effective against viruses (which antibiotics are not), cortisone (for arthritis and other inflammation), and the first psychopharmacological agents (lithium for manic depression, chlorpromazine for schizophrenia). Beta-blockers, anticoagulants, antiarrhythmics, antihistamines, antidepressants, anticonvulsants, steroids, bronchodilators, endocrine regulators, insulin, and cytotoxic drugs against cancers were all available by the 1960s.

Medical practice changed substantially, as one might expect. The bedside manner, formerly prized, was almost deliberately rejected in favor of a businesslike attitude in which the patient was sometimes seen as a veil between the physician and the disease, to be seen through rather than attended to. If there is a pill for every ill, this makes sense; but patients did not always like it. Medical practice, which had always been a private affair, inexorably became a concern of the state in the 20th century, regardless of what character that state had. This was in part because of the needs of the modern state (a healthy workforce), the growing populations of cities, the demands of war, and other factors that have driven or accompanied the general expansion of the role and sphere of the state in the 20th century. But state involvement in medicine was also spurred on in the second part of the century by some shocking disasters of effectiveness, notably thalidomide, which was marketed as a safe sleeping pill despite warnings of side effects being known to the manufacturer, and caused serious fetal defects. Hence the rise of strict standards for clinical trials, of ethics committees, and indeed of the academic profession of bioethics.

European medicine had been touted throughout the empires of European powers, partly because medicine was required in order to achieve imperial ambitions—in the 1880s, an attempt to build the Panama Canal had to be abandoned because the workers simply died of malaria, over 5,000 of them—but also because it provided a justification. Educating the natives in the ways of scientific medicine was surely a good thing. In fact, many of the diseases suffered by the natives had been brought by the settlers, just as the diseases that scourged the settlers were ones absent from their origins. Moreover, the actual curative efficacy of European medicine was on a par with any local tradition, and in some cases probably more harmful.

However, the collapse of political empires in the latter part of the 20th century left a good deal behind, including European medicine—which, by then, had in fact become distinctively successful. Whatever the reasons for it, there is no denying that the middle of the 20th century saw an explosion of curative abilities that is unmatched in history. Thus the persistence of European medicine after the end of European empires is not, I think, a mere symptom of the persistence of colonial power structures after the end of formal colonial occupation. It arises in part from at least the perception of the special curative abilities of European medicine.

Because it has become the standard everywhere, I call the contemporary tradition Mainstream Medicine. It is no longer specifically European. That said, Mainstream Medicine remains distinctly insensitive to the needs of the developing world. Concepts of health and disease differ, and thus so do perceptions of need. In many cases, efforts to intervene on the guilty microorganism are either unethical or ineffective. Consider the effort by several nongovernmental organizations in southern Africa to promote male circumcision as a means to control the spread of HIV/AIDS, which is neither very effective nor ethically sound (in my view). More generally, the development of drugs to take out the guilty microorganism is not the effective or affordable way to tackle diseases whose larger causes are miserable poverty, poor housing and nourishment, social problems such as severe gender inequality, and so forth. Tuberculosis, for example, arises in malnourished populations living in crowded conditions. It is not a highly infectious disease; indeed, living in Johannesburg, a historical home of the disease in Africa, I personally know individuals who have it, yet I do not regard myself at any risk of contracting the disease. A program to eradicate tuberculosis in South Africa would require improving the living conditions of millions of South Africans, and also the effective control of HIV/AIDS, which in turn would require an increase in the social power of young women. The solutions to these problems are clearly not within the purview of Mainstream Medicine, though they perhaps would be within the purview of other medical traditions.

Mainstream Medicine is thus a fascinating object of study. It is wonderfully successful, compared to any other medical tradition, at delivering cures of certain illnesses. But the Golden Age of discovery did not last; since, say, the 1970s, there have been relatively few breakthroughs. This does not mean no improvements: incremental improvements are important, as I will argue in Chapter 5. But there remain very many diseases that Mainstream Medicine cannot really do much about: many cancers, heart disease, many mental illnesses, much diabetes, back pain, and of course minor but persistent ailments like the common cold. There are other ailments, like tuberculosis and HIV/AIDS, where medical intervention is only moderately successful. And, even more frustratingly, there are ailments that Mainstream Medicine *can* treat, but where treatment seems not to solve the larger problem. Malaria is the most obvious example: it is an ancient disease, probably the disease that has claimed the most human lives, and one that we understand and can treat when it arises;

yet we have been unable to push it back despite repeated efforts. And then there are the "neglected diseases," which may be tractable to Mainstream Medicine, but towards which it is not profitable to direct medical research, because those who suffer from them are too poor to pay a profitable price for the outcome.

Against this background, Mainstream Medicine finds itself no less criticized, and perhaps more so, than in the days when the doctor really could do almost nothing. Medical malpractice litigation is common in many places (though much less commonly successful). Even educated persons turn to rival traditions such as homeopathy, attractive for the same reasons it ever was—the suspicion and abhorrence of harmful pharmaceuticals, a suspicion that was probably well founded 300 years ago and, according to some, may still make sense today (we will discuss this in Chapters 5 and 6). Traditions are imported; acupuncture and Ayurvedic medicine are now well established as niche alternatives in the West. Chiropractic and osteopathy are popular given the absence of effective nonsurgical treatment for back pain and the questionable long-term effectiveness of the surgical option. Midwives, who were historically much more interventionist than doctors at a birth (Porter 2002, 119) now promote themselves as taking a more natural, respectful view of childbirth, minimizing intervention and allowing nature to take its course where possible—exactly what the Hippocratic doctor would have said, and would have criticized the midwife for not doing. (The fact that the Mainstream medical profession, gynecology included, remains male-dominated, while midwifery remains almost exclusively female, also plays a role here.) Medicine is the object of study in this book, not Mainstream Medicine alone. However, the influence and ubiquity of Mainstream Medicine clearly requires that it be given special attention, and so it is in this book. Yet that attention is not exclusive. Throughout I make allusions to other traditions, and they form the focus of the last two chapters of the book. Thus I now give a brief survey of Alternative and Traditional Medicine.

1.3 ALTERNATIVE MEDICINE

In Europe, various trades and individuals have long claimed to promote health or heal illness. The formalized medical profession has consistently derided them. In contemporary Europe and America, there is substantial controversy about Alternatives. It is interesting that there was controversy long before Western medicine detached itself from ancient ideas about balancing humors and based itself on anatomical, biological, and chemical knowledge. And even after that re-basing, effective cures remained rare until well into the 20th century. Yet quacks were reviled all along. Thus the contemporary disputes about Alternatives probably have a deeper social explanation than some of the neo-Enlightenment rhetoric might suggest, and certainly have deeper historical roots.

Outside Europe, of course, other medical traditions have existed largely outside the orbit of Western Medicine. It is reasonable to treat Western Alternatives separately from what I should perhaps refer to as *Non-Mainstream Traditions of Medicine* but will instead call *Traditional Medicine*, for convenience. Some of these traditions now form the basis of imported Alternatives in the West. Despite recent blurring, the distinction is reasonable, because the histories are different and deserve their own treatments. I defend this distinction further in Chapter 8.1.

Porter identifies the 18th century as the golden age of Alternatives, due in part to the increasingly commercial character of society. Many Alternative practitioners were extremely entrepreneurial, traveling from town to town, putting on wonderful shows at the market, and riding on before anyone could catch up with them.

Yet "quacks" were not all charlatans; some believed in their cures wholeheartedly, and some may have actually had cures. The most striking example of this was the 19th-century *Eau médicinale*, touted as a cure for gout by a French army officer, Nicolas Husson. Although it was the subject of derision by the Mainstream profession, it was, in fact, effective, because it contained colchicum. Moreover, Mainstream Medicine had no remedy of its own that was effective against gout. Again, this suggests caution in the present day for those who are inclined to dismiss all Alternatives as quackery. If social and professional allegiances could interfere with dispassionate assessments of effectiveness in the 19th century, they can do so in the 21st—indeed, there is good reason to think that they do, as we shall see in Chapters 5 and 6.

Some Alternative Medicine grew decidedly un-quackish in its seriousness, systematicity, and institutional apparatus—although it is hard to find an Alternative that does not base itself in the sage-like insight of some great man. Homeopathy was the "great trail-blazing inspiration" of Alternatives (Porter 2002, 46). It was developed by the German physician Samuel Hahnemann (1755–1833). Homeopathy holds that a vital force underpins all life, and that healing is a matter of stimulating this vital force, which will then defeat the disease. It holds that like cures like, so that to cure one must find a substance that causes symptoms in a healthy human body similar to the ones observed in the diseased body. It also holds that the smaller the dose, the larger the response. These principles are reversals of tenets of Mainstream thought, as we shall discuss in Chapter 8. They may seem radically implausible from a commonsense perspective, but homeopathy retains a strong following, and moreover it is quite possible that Hahnemann's obsession with purity and his distaste for the haphazard concoctions of the regular doctors of his day may actually have meant that, in its time, homeopathic treatment would have done less harm than regular treatment. In places where overdiagnosis and overmedication are contemporary problems, the same could well be true today, even if homeopathy is entirely ineffective—which, as I show in Chapter 8, is not a conclusion we should jump to.

There were a host of less comprehensive systems than Hahnemann's, and many irregulars touted just a single prescription, or specialized in a single disease. Nonetheless, they all tended to expose the deficiencies of Mainstream Medicine as "an obscurantist racket devoted to self-aggrandizement" (Porter 2002, 46). In the 19th century, especially in America, there was much made of man's unnatural way of living, and Mainstream Medicine was portrayed as part of that. Again, by contemporary lights, there is something to this. The 20th century gave rise to epidemiology, which showed that life in large urban conurbations, working practices, diets, and habits were all killing people.

Alternative Medicine thus persisted through the advent of scientific medicine, and it continued in the 20th century, although it declined in the first half of the 20th century with the rise of effective cures (and perhaps due to the ravaging of Europe by war). However, the dip was short-lived, with interest in Alternatives enjoying a rejuvenation in the 1960s. This is partly explained by the anti-authoritarian sentiments of the age, and partly by the fact that, as medicine became more effective, the bedside manner was abandoned. Moreover, Mainstream Medicine became increasingly limited in its scope, ignoring the emotional and decrying the spiritual. Porter writes:

> While people wanted to be relieved and cured, they were also seeking far more from medicine—explanations of their troubles, a sense of wholeness, a key to the problems of life, new feelings, of self-respect and control. If the tenor of orthodox medicine was pessimistic, alternative medicine instilled hope.
>
> (Porter 1997, 51)

Perhaps assisted by widespread economic growth, the popularity of Alternatives has continued to grow. Besides homegrown alternatives, Eastern traditions have been imported to the West: acupuncture, in particular, is a popular therapy originating in China. The menu of Alternatives is huge and various: osteopathy, chiropractic, homeopathy, acupuncture, Reiki healing, Ayurvedic medicine, reflexology, crystal healing, to name a handful. They vary enormously in their philosophical outlook, degrees of naturalism, attitude to Mainstream Medicine, and, of course, recommended treatments. However, when practiced as Alternatives in a Western context, they share a consciousness of the Mainstream tradition, meaning that they position themselves in relation to it in some way, whether by rejecting, augmenting, or extending it. This is inevitable, because Mainstream Medicine cannot be ignored, and every practitioner treating a patient who has chosen to come to him instead of seeking Mainstream treatment must give some minimal indication of why that was a good decision—even if it is as crude as asserting that Western medicine fails to recognize meridian lines, which are part of ancient Chinese thought, as important for health.

Some traditions occupy an ambiguous position, outside the Mainstream medical tradition but more acceptable than the more oppositional varieties. Physiotherapy is perhaps the best example of this, since it does not reject or modify the theoretical basis of Mainstream Medicine and is often used by it (e.g., for postoperative rehabilitation). However, many of its interventions lack a clear Mainstream basis, either in theory or evidence. Massage, dry needling, and even mobility exercises are not part of the Mainstream medical corpus, even if they are not incompatible with it either. And a physiotherapist is far less likely than a doctor to see anything wrong with recommending an arnica gel or lavender-scented heat pack.

There were more *registered* Alternative practitioners than general practitioners in the United Kingdom at the end of the 20th century, and in the United States, 425 million visits per year were paid to Alternative practitioners compared to 325 million to primary care physicians (Porter 2002, 51). For this reason, any philosophical account of medicine must consider Alternatives, and must be able either to incorporate them or to give good reasons for rejecting them.

1.4 NON-WESTERN TRADITIONS

Medical traditions vary in almost every way imaginable. They vary in their degree of naturalism, in their degree of holism, in the way they explain disease, and in the remedies they offer. The ancient Greeks set Western medicine on a naturalistic path (humors were natural— they were literally fluids). For the Western mind, it is tempting to suppose that other traditions are largely non-naturalistic, especially given that the non-Western imports practiced as Alternatives typically include spiritual components. This is not the case. However, non-Western traditions differ from contemporary Mainstream Medicine in that the latter abandoned its classical roots, and based itself on scientific theory, while others did not do so. In the 19th century, many traditions were forced to adapt to Western medical thinking due to the enormous military power of the West. In the 20th century, Mainstream Medicine has exerted a continued force on other traditions as a symbol of modernity, advancement, progress, rationality—and effectiveness.

These forces have led to different interactions between Mainstream and Traditional medicine in different places. In this section I briefly review three traditions and their interactions with Mainstream Medicine. There are many more. My selection of Chinese and Indian medicine is informed by their status as very substantial learned traditions prior to crunching against Mainstream Medicine.

My selection of African medicine is informed partly by local interest, as I live in the region, and partly by the fact that it is quite different from the medicine of the Chinese and Indian traditions, both in its character and in the way it interacted and continues to interact with the Mainstream.

Settlement in India began around the same time as elsewhere, about 10,000 years ago. Early medicine seems to have had a magico-religious basis, similar to the traditions of Mesopotamia and Egypt, with particular deities possessing particular healing powers. The early Indus civilizations seem to have crumbled around 1500 B.C.E. Subsequently, the dominant faith in northern India around 1000 B.C.E. was Veda, a body of teachings owned by a priestly brother-hood of Brahman. There is no distinctive Vedic medicine, however. During the course of the following millennium, various ascetic individuals and sects, notably Buddhists and (what were to become) Jains, introduced new medical practices. The Ayurvedic tradition that came to dominate in the early years of the Common Era traces it-self back to Veda and the Brahman but is very unlikely to actually descend from these. Ayurvedic texts bear very little resemblance to Vedic texts, and much more to Buddhist ones. The earliest Ayurvedic texts originate around 400 C.E., with additions continuing into the 16th century.

Classical Indian medicine was a learned tradition, by which I do not mean to imply superiority to non-learned traditions, but rather a tradition based on written texts and upheld by powerful and long-lasting institutions. In fact, being "learned" in this sense has a price: such traditions tend to revere the past, and to resist change. Of the great Asiatic traditions, Porter writes:

> The consolidation of writing encouraged learned traditions which helped to give permanence to particular corpuses of medical (as well as religious and philosophical) erudition. As

with the writings of Hippocrates and Galen in the West, the result tended to be a glorification of tradition, and the associated belief that a fixed, permanent and perfect medicine had, in a quasi-divine manner, been handed down from some far-distant origin.

(Porter 1997, 135)

Claims of revelation to sages in meditation at the dawn of time are often not borne out by historical evidence, and they do not encourage development.

Ayurvedic medicine consisted in a large volume of guidance on living, eating, exercise (it is related to yoga, now very familiar in the West), as well as a wide variety of pharmaceutical preparations of animal, mineral, and vegetable origin. Surgical techniques are described in one text but appear to have fallen by the wayside subsequently. Texts include detailed anatomical material as well as lengthy philosophical discourses on causation or other matters.

Portuguese settlers arrived in Goa in the first half of the 16th century. Initial warmth quickly declined, and Hindu physicians were effectively banned from Goa after 1600. Later in the 17th century the Dutch East India Company interacted with traditional Indian medicine in reasonably fruitful ways, and initially the British East India Company adopted local medical practices due to the difficulty of importing medications. The English were interested in Indian physic, while the Indians were interested in English surgery. Medical colleges were founded and included Ayurvedic medicine. However, attitudes hardened, ethnocentricity set in, and regulations changed; Ayurvedic medicine was suppressed.

Nationalist independence movements in the 20th century encouraged Indian indigenous medicine. However, Western medicine remains a symbol of progress. This has caused tension. Porter writes:

In recent decades there have been divided loyalties: since independence in 1947, the Indian government has oscillated between commitment to western medicine in the name of progress, and acceptance of the fact that Ayurvedic medicine is widely practised, especially in the countryside, and commands sturdy loyalties.

(Porter 1997, 144)

A central council for Ayurveda was set up in the 1970s and there now exist a number of Ayurvedic (and Yunani, the Islamic medical tradition in India, which I have not discussed here) medical colleges. However, these are sometimes perceived to be second-choice destinations for those unable to make the grade for Mainstream medical school. They receive minimal funding and include some basic education in Mainstream ideas.

In practice, many Ayurvedic practitioners employ Western methods or drugs, including antibiotics. The trend, according to Porter, "is towards the greater assimilation of western medicine," and he points out that Ayurvedic medicine has not gained the same traction in the West that Indian philosophy has (Porter 1997, 146).

Classical Chinese Medicine is likewise a learned tradition, often presented as timeless and quasi-divine in origin. It is sometimes claimed to be unchanged for the past 2,000 years. While this is not quite accurate, it is nonetheless a point of contrast with Western medicine, and reflects the fact that Chinese thought traditionally has not prized novelty.

Chinese medicine is not wholly indigenous, with influences coming from Tibet, India, and central and southeast Asia. Chinese medicine spread wide, and by the 16th century had been or was being introduced to Korea, Japan, and the Philippines, and from the 19th century to the Americas. Acupuncture was practiced in the West from about the 17th century, becoming popular in France during the 19th.

In those areas where it was introduced, Chinese medicine is often still practiced today, as an Alternative alongside Western medicine.

China was unified in the third century B.C.E., and the earliest Chinese medical texts stem from the unifying efforts of the first emperors of the Han Dynasty. Four ancient texts remained authoritative through subsequent developments. There is some suggestion that scholarly interests sometimes prevailed over practical: early descriptions of the tracts along which vital substances flowed bore more relation to anatomical paths of blood vessels than those that followed later, when the theory of systematic correspondence was fully developed (Porter 1997, 153).

Porter emphasizes that Chinese medicine must be understood as a classical system, explaining disease in terms of cosmic factors and influences, rather like Hippocratic thinking in ancient Greece (Porter 1997, 151–152). At the heart of Chinese medical tradition is the notion of *qi*, which permeates the cosmos, and, when accumulates, produces life. *Yin* and *yang* are familiar in a crude form in the West as polar opposites (male/female, hot/cold, etc.). Their relationship is more complex, however; they are ways that *qi* may be distributed. In addition there are *xu wing* or "five phases" (formerly translated as "five elements"): wood, fire, earth, metal, water. The body is a microcosm and its processes are the action of *qi, yin* and *yang*, and the five phases. The majority of medications were pharmaceutical, but acupuncture and moxibustion (the burning of small pellets of wood on the skin) were also used. Both were thought to stimulate the flow of *qi*. Classical Chinese medicine was not religious, and it was naturalistic, but the masses saw ancestors, gods, karma, and sin as causes of disease and sought the help of a range of healers outside the learned tradition, as in the West.

In the 19th century, missionaries arrived from the West, and some saw Chinese medicine as partly responsible for its military defeats. Some sought to modify rather than abandon it; others were more hostile. Largely through foreign influence, medical colleges (teaching Western medicine) were established in the late 19th and early 20th centuries. In the period 1911–1949, Republican China sought to establish a modern, Westernized medical system, and after 1948 the communist regime took the healthcare system over largely unchanged, but started to incorporate Chinese medicine. Chinese and Western medicine came to enjoy equal status, with top physicians expected to know both.

Chinese medicine has changed in the process, with concepts becoming equated to Western ones through translation: for example, *xue* is often translated as *blood*, which is not wrong, but is a substantial simplification (Porter 1997, 161). Nonetheless, knowledge of classical texts remains essential for contemporary Chinese practitioners. In this sense it remains a classical tradition. There is a tendency to seek a synthesis of Western and Chinese medicine. However, Porter points out that there remain fundamental differences:

> Both originally shared common assumptions about the balanced and natural operations of the healthy body and these were inscribed in hallowed texts. Western medicine alone radically broke with this. An entirely new medicine grew up in the West— scientific medicine—building upon the new sorts of knowledge, programme and power that followed from anatomy and the investigations of the body it opened.
>
> (Porter 1997, 162)

Chinese medicine has enjoyed recent popularity in the West, notably acupuncture; but as Porter points out this is in part because of scientific explanations for its effectiveness (in terms of endorphins), as opposed to acceptance on its own terms (meridian lines through which *qi* flows). It is thus not clear whether Chinese medicine will merge with Western. The rise of China as an economic force is also an important factor, since, as these narratives show, the spread of a medical tradition has at least as much to do with the political muscle behind it as its curative potency.

In South Africa and southern African more generally, a vibrant traditional healing practice persists. The tradition is not learned, in the sense given above: it is not text-based, but passed on orally and through training processes. It is primarily non-natural, with diseases being caused by ancestors, malignant spirits, curses, or similar.

In some areas, African medicine is the only medicine available. However, this is not the whole explanation for its persistence and popularity. Well-heeled patients may still consult a *sangoma*, or may consult both a Mainstream doctor and a *sangoma*. Estimates suggest that anything from 60% to 80% of the population consult or have consulted *sangomas*, despite the fact that the tradition sits uneasily with conservative African Christianity, is derided by Mainstream Medicine, and, arguably, is seriously misrepresented in legislation attempting to regulate it (Thornton 2009). Moreover, this is not poor man's medicine: *sangomas* may charge comparable fees to Mainstream doctors; and indeed many people will consult *sangomas* in conjunction with consulting a Mainstream doctor.

All of this is hard to explain if *sangomas* are thought of as a sort of African version of a Mainstream doctor, with the same set of goals but a different set of tools. Indeed, it appears that this is not the right way to think about *sangomas*. As Robert Thornton puts it, "*sangomas* . . . are neither particularly traditional nor healers as these terms are usually used" (2009, 31). Their practice is a live one that cannot be "relegated to the past" (2009, 32); and it is moreover not solely concerned with healing, as that is understood in Mainstream Medicine.

There are a number of different specialties or practices within the traditional medicine of the region. The classic treatment is Harriet Ngubane's book on Zulu medicine (Ngubane 1977), which is a rare account of practices that otherwise tend to remain even more deeply hidden from scholarly view than the medical encounters of text-based traditions. However, in what follows I rely on a broader, shorter, and more recent survey by Robert Thornton.

Here is Thornton's account of divination, which is one kind of traditional African medical diagnostic practice:

> This practice involves the release from cupped hands of a set of objects (*tinhlolo*) onto a grass mat that is situated between the diviner and the client.... When these are thrown onto the mat, the objects land in a configuration that is "read" through a rhythmic verbal interaction between client and healer concerning the meanings of the *tinhlolo*. A diagnosis or possible solution to the problem that is being addressed gradually emerges through the interaction between client, healer and the pattern of the objects.
>
> (Thornton 2009, 24–25)

Pharmacology seems to play a lesser role in African therapeutics than in European, Indian, or Chinese traditions. Thornton says that clients expect to receive "a diagnosis, advice, or herbal remedies as a result" of a divination (Thornton 2009, 24), meaning that something like a drug is by no means the only or usual prescription. Even when an herbal remedy is prescribed, it is not typically intended to be pharmacologically active, as one would assume from the perspective of Mainstream Medicine:

> Most of these [herbal remedies] are not used as pharmacological agents, but rather used in a ritual or for steam or smoke baths, inhaled as smoke or steam, applied as rubs, or worn as amulets. Herbs may be ingested orally, vaginally, anally via enemas, or through small cuts in the skin, but whatever the pharmacological activity the original herb might have (or might have had) is

often not the goal or rationale of the treatment. Since there is no standardization of collection, drying, storage, or other treatment of herbs and other "medicinal" products, much of their chemical activity is lost, modified, or otherwise transformed. In any case, this is rarely the point.

(Thornton 2009, 25)

Thornton's account brings out the crassness of simply assuming that the *sangoma* is an African version of the Mainstream medic with a different set of tools. It also brings out the centrality of understanding in the tradition of the *sangoma*. Thornton argues that legislation wrongly restricts the *sangoma* to weak medicine and social work, when "This is certainly not the vision of most healers, who seem to understand themselves as belonging to an intellectual tradition of which healing is just one part" (Thornton 2009, 23).

Traditional African medicine occupies a complex position in South African society. It does not enjoy high public status, being associated with superstition at best, and horrific crimes at worst. Newspapers regularly carry stories of the imprisoning of albinos for the harvesting of body parts, the kidnapping of children for similar purposes, recommending outlandish remedies for HIV such as sleeping with a virgin, and so forth. Assessing the veracity of these stories is difficult, but there is little doubt that some very unpleasant things go on. There is a dark side to the body of thought and practice that also gives rise to African medicine: black magic ("voodoo") is a live practice in the region, although even harder to depict in a scholarly way than curative practices. All of this flies in the face of the progressive ideals of the region, as well as the conservative forces of evangelical Christianity, Catholicism, Calvinist Protestantism, and those few who hanker after the "old days" of apartheid.

Yet despite these intense pressures, and the tension with modernizing aspirations, the tradition persists and evolves: for instance, *sangomas* use cellphones to run their businesses. Strong feelings of cultural identity explain part of this persistence. There are, however, a handful of White *sangomas*, perhaps suggesting that the tradition is less parochial and more universal in its conceptual scheme than the centrality of local ancestors might make it seem. It will be fascinating to see how the tradition develops—whether it becomes formalized and professionalized in the Western sense; whether more writings arise; whether it instead adopts a reactionary traditionalism; or whether it simply dies out. In some ways, interaction between African and Western medicine has been remarkably minimal to date, because of the apartheid system, which took little interest in what Blacks did among themselves. Since the end of apartheid, however, high-status people will consult *sangomas*, or even be *sangomas* themselves—but typically behind closed doors, rather as a Western leader might quietly consult a priest or practice Buddhist meditation. African medicine has existed below the radar, but it does not seem plausible that this can continue. The real encounter between Mainstream and African medicine may only now be beginning.

1.5 CONCLUSION

Does this extremely incomplete survey of various historical and contemporary medical traditions tell us what medicine is? It tells us something: that medicine involves health and disease, and the attempt to promote the former and thwart the latter. But it also tells us that medicine is not homogenous. It is a radically diverse bundle of activities, even within a given tradition; and different traditions have quite different characters. The question then arises: what is it that makes all of these activities and traditions identifiably medical?

One obvious answer to this question is that they have a shared goal, that of promoting health and healing sickness. Unfortunately this is not enough, since not any well-meaning effort to promote health or cure disease counts as medicine—otherwise I would be a doctor merely because in winter I ensure my children's bedrooms are not so cold as to induce hypothermia. Medicine is a field of *experts*. But experts in what, given that various medical traditions and practices share no common activity?

Answering this question will help us understand what medicine is; and this is the sort of question that a philosopher can hope to help with. Whatever else we do, philosophers are specialists at looking for principles: patterns in the mess of the mundane. The sheer diversity of medical practice is what makes philosophy of medicine a necessary accompaniment to history, sociology, and anthropology in gaining a real understanding of medicine. We cannot decide what attitude to adopt toward medicine without knowing what it is, and we cannot say what medicine is merely by looking at medicine.

Chapter 2

The Goal of Medicine

2.1 HOW TO ANSWER THE QUESTION

An empirical survey will give us one kind of answer to the question "What is medicine?"—but not a very satisfactory one.[1] It will not tell us what these various medical traditions have in common, nor what makes them medicine. It will not tell us how medicine differs from carpentry or law; it will simply enumerate the differences. If we want the kind of answer that would help us understand medicine, then we need to elaborate the "What is . . ." question.

I propose three sub-questions that, I suggest, any adequate account of medicine must answer.

1. What is (are) the goal(s) of medicine?
2. What is the core business of medicine?
3. What are health and disease?

---

1. This chapter draws heavily on my inaugural lecture at the University of Johannesburg, "Prediction and Medicine," delivered in 2016, as well as the replies by Chadwin Harris and Thaddeus Metz. It also draws upon the subsequent paper, responses, and author's reply published in the *Journal of Philosophy of Medicine* (Broadbent 2018a; Broadbent 2018b; Harris 2018; Metz 2018).

If you can provide a good answer to each of these questions, then you are well on your way to understanding what medicine is. I will now explain why.

The first thing an account of medicine calls for is an account of the goal or goals of medicine. Ascribing goals to a complex social institution is always difficult because the various participants may have all kinds of objectives. We might want to say that law aims to achieve predictable and procedurally fair settlements of disputes, but it may be that rather few actual lawyers or litigants seek a fair settlement, and it may be that fair settlements are rarely delivered. It is very natural to say that the goal of medicine is to heal the sick; but there are many instances of medical practice that are not meant to be curative (such as cosmetic surgery or pain relief), and many that do not in fact cure. An account of the nature of medicine needs to find a way to avoid getting bogged down in this wider problem, as well as identifying the goal or goals of medicine, or indeed arguing that there are none.

Second, an account of medicine requires saying not only what its goals are, but what its core business is: what it actually does. It would be odd if goals and activities were not related, but there is no reason to assume that the activities can be "read off" the goals. This is particularly important in medicine because, as I say repeatedly in this book, it is quite reasonable to accept that the goal of medicine is to heal the sick, and yet equally reasonable to take the view that this is not what medicine mostly does.

There is no easy response to be had by asserting that the core business of medicine is *trying* to heal the sick. Normally, one's core business is something one actually does, not something one merely tries to do. Blacksmiths do more than try to make horseshoes, and taxi drivers do more than try to take you places. The core business of a profession or tradition is related to a *competence* or skill of some kind that its practitioners have. Maybe doctors try harder than everyone else, but at the very least, it seems important that we open the door for other accounts of what doctors are good at, besides trying. The mere goal of healing a sick person is something that any nice person might share, from the neurosurgeon to the kindly relative who brings a cup of tea, mops the brow, or perhaps misguidedly applies a home remedy. Medicine is often professionalized, and fiercely so in the contemporary world. Assuming that professionals have specific competences relating to their professional field, we should be able to say what they are. We can hardly assume that the competence is identical with obtaining the goal of the profession, any more than we can assume that we will always get what we want.

Third, medicine is clearly bound up with notions of health and disease. Thus, to be complete, any account of medicine requires an account of the nature of health and disease. This has seemed obvious to everyone working on the philosophy of medicine, because everyone agrees that medicine centrally involves health and disease, even if they disagree on the nature of that involvement.

2.2 THE CURATIVE THESIS

In all times and in all places, medicine has had something to do with health, disease, and the transition from one to the other. A natural reaction is to think that *medicine is the sustained and organized effort to heal the sick, or prevent them getting sick in the first place.* I call this view the *Curative Thesis.* This is a view both about the goal of medicine, and its core business. Both are to heal the sick—to cure. Hence the name. Prevention is a natural companion, since it is also a way of bringing it about that there is health rather than disease. But cure comes first, because people need to have suffered from diseases before the idea of preventing them comes up, and when people suffer from diseases the first step is to try to heal them. The place of prevention is medicine is complex and not uncontested, and I will discuss the relationship between prevention and cure further in section 2.4.

To understand the Curative Thesis, we need to better understand the nature of cure. It is important not to assume that cure is what medicine aims at, and that whatever medicine aims at is cure. Even if one accepts the Curative Thesis, one cannot *define* cure by its role as the goal of medicine, since that would render the thesis tautological and empty. The content of the Curative Thesis depends on what is meant by "cure," which must therefore be decided by anyone endorsing that thesis, as well as by anyone rejecting it.

It is helpful to distinguish between three related ideas: *cure, therapy*, and *medicine*. The latter is the wider profession, within which the former two clearly play some role. I suggest that *cure* be understood as an intervention that removes a disease. (I use "heal" and "cure" interchangeably.) I suggest that *therapy* be understood as any intervention that alleviates the suffering or harm caused by disease, but does not necessarily remove it. Defined this way, therapy might be broadly construed to include cure (but not vice versa); however, for clarity, it is probably best to use "therapy" narrowly in a way that excludes cases of complete removal of the disease, and keep the categories of cure and therapy separate. The ambiguous term "treatment" might be understood as encompassing both therapy (narrowly construed) and cure.

These rough categories may not be applicable universally, but they seem to be present in the Hippocratic tradition. Consider this definition of medicine from *The Science of Medicine*, as:

> the compete removal of distress of the sick, the alleviation of the more violent diseases, and the refusal to undertake to cure cases in which the disease has already won the mastery, knowing that everything is not possible to medicine.
>
> (Miles 2005)

This passage seems to indicate a distinction between complete removal and alleviation, corresponding to my distinction between cure and therapy. It also indicates the Hippocratic suspicion of wild promise of cure, considered the preserve of the quack.

In contemporary medicine, many treatments are therapies rather than cures. For example, antiretrovirals for HIV/AIDS allow the patient to live a much longer and healthier life than she otherwise would, but they do not remove the disease. Insulin for diabetes is another example of a therapy, extending across the course of life, but not removing the disease. Insulin supplies what the body needs to function, but does not mend the broken pancreas, which normally supplies the body with insulin. Epileptics can have their epilepsy "controlled" but not cured: if the supply of medication is disrupted, the controlled epilepsy will be uncontrolled, and the epileptic will probably suffer seizures.

The distinction between cure and therapy may not always be clear. One factor is whether the treatment is lifelong or episodic: can something be considered a cure if it must be administered consistently for the rest of the patient's life? I would be inclined to say "no," because the underlying cause is not dealt with, and the patient remains dependent on the cure. It may be difficult to draw a principled line here. What do we say of an implant that needs renewing every ten years? What would we say about a course of treatment that stops eventually, but after a long period of time—say ten years? What about a course of treatment that ends not at death but at another life stage such as puberty or menopause? I am not sure what to say about these cases. But nor am I sure that this matters. Lack of clarity at the border does not obviate the value of the distinction, which is clear in some cases even if not in others.

Another factor in distinguishing therapy and cure is whether the removal of symptoms is sufficient for the removal of the disease. Some of our diseases are defined symptomatically, especially mental ones, but not only them. Whether one sees a disease as cured or merely alleviated may depend on whether we think it is anything beyond a collection of symptoms. For example, whether antidepressants cure or remove depression depends partly on what you think depression is. If it is merely a collection of symptoms (feeling down, losing sleep, lacking motivation), then an effective antidepressant could cure it by addressing the symptoms. If, on the other hand, it is a malfunction of the brain, then antidepressants might be the equivalent of insulin, supplying what is needed to permit normal function, but not curing the disease in the sense of, as it were, mending the broken brain.

Given this general distinction between therapy and cure, there is still some scope for varying the strictness of the definition of "cure." A *strict cure* might be *an intervention that completely removes a disease and renders the patient as well as if they had never suffered from it.* If I adopt the strict notion of cure, refuting the Curative Thesis will be easy; even our best cures will rarely measure up. The interest of such an implausible Curative Thesis would be quite limited. At the other extreme, I could adopt a very *relaxed definition of cure,* such that *an intervention that is in any way effective at alleviating the harm or suffering caused by a disease* counts as a cure. Depending on what "effective" means, this could include supportive treatment: mopping the brow, administering sips of water, and so forth. This definition of cure would basically expand it to include the whole of what I called "therapy" above, and effectively abandon the distinction between cure and therapy. It could even go further than what I called therapy, and also include sound advice: rest for a couple of days and you will be fine. Or indeed lifestyle advice: try losing a bit of weight, taking some exercise, eating regular meals, getting a bit more sleep; and so forth. It may even be that the comfort of knowing what you have somewhat alleviates the suffering caused by a disease, in some cases. If any situation where the doctor takes or advises an action that is for the better of the patient will count as a cure, the Curative Thesis is very hard to refute. But the interest of such a weak Curative Thesis would be just as limited as the interest of the implausibly strong version.

An interesting version of the Curative Thesis must use a *moderate definition of cure*, such that a cure is *an intervention that is reasonably effective at altering the course of a disease for the better*. Thus "at all effective" in the relaxed definition is replaced by "reasonably effective." Clearly there are many degrees between the two poles I have indicated. My moderate definition is not the only one possible, but I think it is reasonable. It employs the notion of reasonable effectiveness, which is vague but important (and which we will consider further in Chapter 5). The term "reasonable" can be fleshed out in a professional practice so as to receive an objective and case-based definition: it is used this way in law. Whatever more detailed definition it might receive, I mean it to exclude "sort your life out" advice, and mopping of the brow. It also excludes pain relief, which alleviates suffering but does not improve the course of the disease, by shortening its duration or improving the chances of recovery. (Of course, to the extent that a pain relief agent does these things, it may be curative, as it appears that aspirin may be in some respects.) This accords with common usage; it is hard to defend the claim that pain relief, and indeed the alleviation of suffering more generally, are kinds of cure, when common usage clearly differentiates pain relief from cure.

Yet it is not too strict: a person does not have to be put back in the position they would have been if they had not had the disease, or "wholly cured". They may still be unwell after the administration of a cure in the moderate sense, but it is still a cure if it is reasonably effective at making them better than they would have been without it. (I intend the interpretation of the vagueness in "reasonably effective" to cover the question of how much improvement is necessary before an intervention counts as a cure; I take it that very small improvements would not warrant the use of the term "reasonably effective" in this context, nor the term "cure.")

I suspect that the Curative Thesis, or something similar, is widely held. For example, Jacob Stegenga defines Medical Nihilism as "the view that we should have little confidence in the effectiveness of medical interventions" (Stegenga 2018, 1), which only makes sense if effective intervention is essential to medicine, since otherwise loss of faith in effective intervention would mean only *therapeutic* nihilism (Porter 2002, 39), and not nihilism about medicine as a whole. (We will discuss this further in Chapter 6.) More generally, effectiveness is a watchword of contemporary thought about medicine. The Evidence-Based Medicine movement focuses entirely on evidence concerning effectiveness of interventions (Sackett and Rosenberg 1995; Evans et al. 2011). Critiques of homeopathy and other alternatives often turn on the lack of evidence for their effectiveness (Hansen and Klemmens 2016). Effectiveness is the order of the day, and in particular, lack of effectiveness sounds the death knell for a practice hoping to be called medical, in the eyes of many people.

2.3 WHAT DOES "GOAL OF MEDICINE" MEAN?

The Curative Thesis says that medicine is a sustained and organized effort to heal the sick (and stop people getting sick in the first place, which I discuss in the next section). The word "effort" implies that healing the sick is a goal or purpose of medical activity. Before we assess the Curative Thesis, we need to clarify what it is for medicine to have cure as a goal. What does it mean to assert that the goal of medicine is to cure? Only once we have decided what it means can we ask if it is true.

It is no easy matter to say what words like "goal," "aim," purpose," "point," and so forth mean as they apply to medicine. Obviously not every single action by a doctor has the goal of cure. Nor does every patient who consults a doctor want cure. Cosmetic surgery, assisted suicide, pain relief, temporary alleviation of grief after a loss so as to get some sleep, circumcision, even something to help a student focus on her studies, are all reasons that a person might seek medical assistance or advice, and without shoehorning it is hard to make them fit the template of cure.

One could argue that some or all such consultations or practices are illegitimate. This is not plausible, but even if it were, the purposes of agents in medical activities and interactions could be hugely various, and need not prioritize or even include healing the sick. The parent might take the child to the doctor because the cough is annoying the parent and not because the parent wants the child to be healthy. The doctor might prescribe something with the purpose of getting the miserable afternoon of consultations over with as soon as possible. Individual doctors have all kinds of motivations: to profit, seduce, get home in time for a favorite television program, or whatever. Individual patients may seek all kinds of things, including reassurance, better sleep, a modification of their bodies to more closely resemble the bodies of the opposite sex, and so forth.

Thus to say that medicine has a goal or purpose is to assert something quite vague. However, as anyone who spends time with philosophy eventually realizes, there is often a tradeoff between how easy it is to say something precisely, and how important it is to say it. To say that medicine has cure as a goal means several things. First, it means that a substantial number of medical activities and interactions have cure as the *ostensible* goal of all the parties. Some doctors might lie, and be motivated solely by profit; but it is enough that they *pretend* to have cure as the goal. Patients might likewise lie (harder to see why, but conceivable). But in a substantial number of medical activities and interactions, cure must be, on the face of things, a shared goal of the parties to the interaction. Second, society at large must have cure as a goal for medicine. That is, society at large must make its support for medicine (legal, economic, political, social) conditional on medical professionals having cure as the goal of a substantial number of their professional activities.

The words "substantial number" in both parts of this definition permit doctors to carry on activities with other purposes in mind besides cure, so long as they remain "substantially" focused on cure. Just how focused on cure they must be is, I think, a matter of discussion and negotiation. How focused on cure they in fact are may vary between traditions, or disciplines within a tradition, or even individual practitioners. Some orthodontists probably cure nothing in their entire lives, even though there is a case to be made that orthodontists can devote themselves to cure (perhaps after an injury that would otherwise leave the patient unable to eat) rather than enforcing a standard smile. Anesthetists also cure nothing, though their social role is morally much more defensible than that of orthodontists, since they alleviate rather than cause pain. Societies form their own views on these matters, and so it is right to leave "substantial number" vague.

2.4 IS CURE THE GOAL OF MEDICINE?

In the sense just given, it seems to me very hard to dispute that cure is at least *a* goal of medicine. Show me a medical tradition that does not have cure as a goal—that does not have cure as the ostensible intention of the parties to a substantial number of medical activities and interactions, or of which the embedding society does not expect that its practitioners will have cure as their goal in a substantial number of such activities and interactions—and I will change my mind. I cannot think of one.

The Curative Thesis implies more than this, however. The Curative Thesis asserts that medicine is the sustained and organized effort to heal the sick. It follows from this that cure is not only a goal of medicine, but *the goal*. This follows on the assumption that medicine cannot have a goal without making a sustained and organized effort to reach it. I think this is a reasonable assumption; something one merely wants but makes no real effort to get is not a goal, in this sense, but a hope. Since no other efforts are mentioned, there are no other goals.

This strong thesis is compatible with more than you might think. It is compatible with doctors, patients, and others in the medical activity or interaction having other goals at times, according to the analysis I have just offered of what it means for medicine to have cure as a goal. The Curative Thesis does not interfere with the "substantial number" clause. Rather, it asserts that there are no *other* goals that meet the definition I have offered.

Further, it allows for a hierarchy of purposes. Much medicine involves diagnosis and prognosis, a point that I will emphasize at length in the next chapter. These are often goals of medical activities and interactions. In asserting that cure is *the* goal of medicine, the Curative Thesis need not deny this. It need only assert that these goals are subservient to the overarching goal of cure, either of this patient, or of others.

I have not found anyone who denies that cure is at least *a* goal of medicine. However, the Curative Thesis disagrees with those who have proposed lists of goals for medicine. Where the goals of medicine have been thought about, the result has usually been a list. (The most recent such list, at present, is provided by Christopher Boorse.) This raises a question: is cure (with prevention) the only goal of medicine, or are there a plurality of goals? In particular, are there goals that are *independent* of cure (or prevention), or do the various goals all *serve* cure (or prevention)?

It might seem that the reasonable position to take is pluralism: medicine has many goals, cure among them. However, this proves much less defensible on inspection than one might imagine.

I believe that the Curative Thesis is correct that cure is not just a goal of medicine but *the* goal.

Boorse's list covers many of the same ingredients as other lists, and so I propose to work through it, and show how none of these goals is both independent of medicine, and really a goal of medicine in general, rather than a use that a society may make of medicine, or may not, depending on its value system.

Boorse's list of goals is as follows.

Goals of Benefit to the Patient

 I. Preventing pathological conditions
 II. Reducing the severity of pathological conditions
 III. Ameliorating the effects of pathological conditions
 IV. Using biomedical knowledge or technology in the best interests of the patient

Knowledge Goals

 V. Discovering the diagnosis, etiology, and prognosis of the patient's disease, including its response to various treatments
 VI. Gaining scientific knowledge about the patient's disease type and disease in general, including their response to various treatments
 VII. Gaining scientific knowledge of normal bodily function
  (Boorse 2016, 170)

Boorse says that his list is comprehensive, but limited to individual patient care.

In my view, pain relief (Boorse's goal III) is a use of medicine but not a goal, while the goal of prevention (Boorse's goal I) is part of the same goal as the goal of cure. Below I devote a section each to exploring the places of pain relief and prevention respectively (sections 2.5 and 2.6).

My response to the remaining plurality of goals is that none of these are goals of medicine, except through their relationship with cure.

This is easiest to see in relation to the Knowledge Goals (V–VII). A mere inquiry, for the sake of interest, into human health would be quite different from actual medical science. This is true whether the inquiry is carried on in a laboratory, in a population-level study, or in a clinical setting. Biological inquiry into the human body does occur, and has occurred. But to the extent that it can be described as medical science, it has the ultimate goal—however remote—of serving our ability to cure. Once again, the situation is complicated by the fact that particular individuals may be motivated by curiosity, others by greed, others by the desire for glory, etc. But the *ostensible* goal, and the goal society expects such work to have, is to serve cure. This service may be remote and its form may not be known; William Harvey's publications on the circulation of the blood did not yield immediate therapeutic benefits. Yet they were not understood as having merely intellectual interest, as would a description of the circulation of sap in trees. Indeed, it was a cause of nihilistic sentiments in European medicine that the steady accumulation of medical knowledge from the Renaissance through the 19th century failed to produce therapeutic breakthrough. This makes no sense if cure was not the goal, in the sense I have given, of the pursuit of that knowledge.

Boorse also lists (as goal IV) the use of biomedical knowledge or technology in the best interests of the patient. However, this is not a goal of medicine, but a goal that a certain society and cultural perspective has for the use of medicine. Not all traditions are naturalistic, and not all involve biomedical knowledge and technology. Moreover, not all cultures see the individual as the locus of moral worth. A culture that saw the family unit or community as the locus of moral worth would not necessarily subscribe to this view, but could surely have medicine nonetheless. And in fact they do. Thaddeus Metz points out that the southern African stance of *Ubuntu* might yield quite different consequences about patient confidentiality and about treatment decisions than contemporary Western bioethics (Metz and Gaie 2010, 278–9). Medicine seeks to cure sickness, but why it does so, and whose interests it serves in doing so, are separate matters.

2.5 MEDICINE AND PAIN

Pain relief (falling under Boorse's goal III) is not cure, by my definition or common usage. Is it a goal of medicine? In my view, the correct answer is that it is a *use* of medicine and medical knowledge, but not a goal of medicine per se.[2]

There is no doubt that pain relief is highly prized by, and expected of, contemporary Mainstream Medicine. However, this tradition inherits the values of a society that attaches a strong disvalue to pain. Not all societies do, historically or geographically. And not all medicine has, or does, place such an emphasis on alleviating discomfort as contemporary Mainstream Medicine.

Even within Mainstream Medicine, the use of pain medication varies widely from place to place, culture to culture, and even within cultures depending on the status of the sufferer (the historian of medicine and bioethicist Daniel Goldberg tells me in conversation that if you are white, male, and middle-aged, relieving your pain is seen as more important). Its use is not always seen as medically helpful: too much pain relief can make diagnosis more difficult. This was true in

2. I am indebted to Daniel Goldberg throughout this section.

the past and remains true in Mainstream Medicine today, for example
in emergency rooms where over-hasty administration of opioids is
avoided for this reason. In the 17th, 18th, and 19th centuries pain
would have been seen as indicating vital force, and thus not necessarily a bad sign; its absence, on the other hand, might be taken as
indicating the imminence of death, and often correctly so. In such
contexts pain relief is not the primary goal, and perhaps not even a
goal at all.

Moreover, many medical procedures are painful, even within
Mainstream Medicine. Even injections or the insertion of a drip
are often somewhat painful. Pain is caused by physiotherapy, chemotherapy, and almost all surgery. I conducted a quick social
media survey of my friends, who are primarily privileged people
living in developed countries, and received a list of painful medical experiences: refusal and/or inability to give further pain relief
after knee surgery once the "maximum dose" of morphine had been
reached despite severe continuing pain; cortisone injection into
the ball of the big toe, without anesthetic; bone lengthening; en-
doscopy and colonoscopy; removal of pain relief in the case of an
uncertain diagnosis and instead the prescription of antidepressants;
vaccinations; the strong recommendation against almost all
pain medication during pregnancy and childbirth and during
breastfeeding; bone marrow extraction; and taking spinal fluid. We
expect doctors to seek to minimize and mitigate the pain they cause,
as well as help us with pain they do not cause. But this does not make
pain relief a goal of medicine, only a use of it. And if there is a conflict
between cure and pain relief, cure will win more than one might imagine. I have seen people intubated in intensive care; the only way to
overcome the very strong urge to pull the tube from the throat is the
tie the patient's hands to the bed. This is extremely unpleasant, and
few would willingly submit to it. Sedation can be used to ensure that
few remember it either. It is very useful for Mainstream Medicine to

have the reputation of being painless, but this usefulness also serves the purpose of pursuing cure, not the purpose of pursuing pain relief.

If we suffered pain but no disease, we might have a profession dedicated to alleviating pain; but it would not be recognizable as medicine. Whereas if we suffered disease but no pain, the effort to deal with disease would have much in common with actual medicine. In the real world, where we suffer both disease and pain, the effort to deal with disease is medicine, and the effort to deal with pain accompanies medicine to a greater or lesser extent depending on social and cultural attitudes.

Even setting aside the existence of places and times where pain relief has not been prioritized, Mainstream Medicine itself often involves considerable pain. It is a myth that medical pain is either a thing of the past, or confined to the uncivilized portion of the present. Pain relief is a use of medical skills and tools, but not the ultimate goal of medicine. Perhaps it "kicks in" when cure is hopeless, and certainly in our day and age the minimization of pain during medical treatment is given high priority; but even in our current age of extreme pain avoidance, the attempt to relieve pain never takes priority over the pursuit of cure in the practice of medicine.

In sum: medicine does not always seek to relieve pain; medicine often causes pain and proceeds nonetheless. Taking either of these as a major premise, and inserting a premise to the effect that nothing can be the ultimate goal of medicine if some other goal ever takes precedence, and further that pursuit of the goal of cure does sometimes take precedence, we have the conclusion that pain relief is not the goal of medicine.

One might object that the same argument could be turned against the notion that cure is the goal of medicine. Do we not sometimes forgo the seeking of cure in order to avoid pain? Consider the days of surgery before anesthetic: many procedures were avoided because they were too painful. Consider the person (like myself) who hates injections and avoids them whenever possible, including some situations where the injection would be medically indicated. Consider palliative care, where relief of pain can take precedence over the prolonging of life, to the extent that in a fair number of jurisdictions where euthanasia is illegal it is nonetheless legal to knowingly administer a lethal or dangerous dose of morphine provided that the objective is pain relief and not death. Are not these cases where medicine is practiced, but where the goal of pain relief is prioritized over the goal of cure?

My response is that medicine is not practiced in any of these cases. This is obvious where surgery is avoided because of pain. Clearly, there is no medicine taking place; what is avoided is medical treatment, on the basis that it is too painful. This is not medicine pursuing a goal other than cure; it is a potential patient refusing medical treatment because he, and not medicine, prioritizes the goal of avoiding pain. Perhaps he doesn't even prioritize that goal; he wishes he could stand the pain, but fear stops him. This is certainly true for people like myself who dislike needles and avoid medically indicated treatments as a consequence. Indeed, the way I described this case used the words "medically indicated." Perhaps I was cheating, but in honesty I thought of this phrase before considering that it supported my argument, and decided against changing it because it is the most exact way to express that aspect of the situation. A doctor might reason with the needle-phobic, urging her that she needs the treatment, or the drip; and this would be on medical grounds. Clearly, someone avoiding an injection or drip is avoiding medical treatment for her own reasons, not prioritizing that avoidance in pursuit of medical goals.

Palliative care provides a more compelling counterexample to my assertion that medicine never prioritizes pain relief over cure. It uses medical techniques, is executed by medical professionals in specifically designated institutions that form part of the larger network of medical institutions (alongside hospitals, clinics, GPs' practices, laboratories, and so forth), and generally "looks medical." What reason could there be to deny that this is medicine? Surely the only reason we could have to look for such a reason would be a strained insistence on the part of an unhappy author who has committed himself to an otherwise untenable theory that forms an essential part of his book?

My response is that palliative care is consistent with my assertion that pain relief is a use of medical skills and tools, but not a goal of medicine. Palliative care is where people go to die comfortably. It is not where doctors aim to send their patients. It is where they send them when cure is no longer deemed possible. Palliative care is not a case of medicine aiming at pain relief above cure. Even if there are cases where palliative care is chosen over the attempt to cure, this is not for medical reasons. I might forgo chemotherapy on the basis that, while it will extend my life, it will not cure me (given the specific cancer I have), and that it will reduce my quality of life horrendously in the extra period of life I do gain. I might rather choose to accept my fate and in so doing aim at achieving a deeper kind of understanding and peace, of the sort we all hope to achieve before we leave this life. This, again, is not medicine seeking pain relief rather than cure; it is an individual making a personal and possibly spiritual decision. It is rarely taken in cases where cure is a viable goal, usually confined to cases where cure is very uncertain, but pain and suffering are certain.

It is, therefore, not as easy as it seems to find a case where medicine seeks pain relief over cure. The individual may forgo cure because it is too painful, but even this is unusual in the case of life-threatening illness where cure is a real possibility—and even in cases where cure is very uncertain, or where life will be prolonged by a short period only, as is often the case with chemotherapy. Voluntary euthanasia is a case where pain relief is sought at the expense of life, but no doctor will recommend this as a medical treatment. They may conceivable advise it on the basis of the probable future suffering of the patient and probable lack of cure (although they risk going to prison for doing so) but in doing so, it is clear that they are using their medical knowledge of the course of the disease for a purpose that is either non-medical, or at least that is secondary to the goal of cure, and that only "kicks in" when there is no hope of cure.

2.6 PREVENTION, CURE, AND MEDICINE

The place of prevention in medicine is complicated. In contemporary Mainstream Medicine, and in its recent past, a lot of emphasis has been laid on prevention. The rise of "preventive medicine" is driven by the obvious idea that, in general, if people do not get sick in the first place, that probably serves the purpose of curing people better than allowing them to get sick and then curing them. (This won't always be true: there might be some value in getting sick first, in some cases, e.g. immunity. But those are exceptions to the rule.) Vaccination is one of the wonders of contemporary Mainstream Medicine; the elimination of smallpox is due, not to antibiotics alone, but to vaccination. The Hippocratic tradition likewise encouraged lifestyle choices (as we would call them today) that would aid prevention, perhaps because it was conscious that its curative powers were limited. African healers will advise on the location of villages and buildings, and on precautions to appease ancestors or ward off ill-wishers, all with a view to preventing disease. Chinese and Indian Medicine both include substantial bodies of advice on how to eat, exercise, and so forth. There is no doubt that medicine involves itself heavily with prevention.

On the other hand, preventing disease involves a lot more than medicine, and so if prevention is a goal of medicine, it seems to swell the scope of medicine so that it encompasses much more than medical professionals typically regard as their domain. Doctors do not design towns or menus or chairs. They may be *consulted* in these things, if health is considered. But they are not charged with doing these things. It is not their professional domain.

What is more, if prevention is a goal of medicine, then medicine becomes political. Almost any decision can impact health. The goal of preventing health may thus come into conflict with other goals that may be operating in these decisions, and where there is conflict of goals, there is politics. Town planning, working hours, diet, income levels and disparities, social conventions, cultural norms and taboos—these all have an effect on health. If medicine were concerned with health prevention, would it not be concerned with all these things?

There *are* those such as Sir Michael Marmot who argue that these things are all the concern of medicine, since they are all determinants of health. In one sense, there is no doubt that he is right: the medical does have an interest in almost anything. And indeed the relevant professionals may be consulted in almost anything, from town planning to the design of airplane seats, with a view to preventing disease. However, when Marmot suggests that physicians have a duty to concern themselves with socioeconomic inequality because it causes ill health (Marmot 2006), he is probably going further than most medical professionals would want to accompany him. Most medical professionals do not agree with Marmot's assertion that medicine is professionally concerned with intervening on all causes of ill-health. Otherwise the medics in attendance at a boxing match would consider themselves professionally obliged to intervene to stop the fight.

There is a lot to say about prevention and medicine. Effective prevention often involves epidemiology, to discover causes of disease and assess effectiveness of interventions, and the interventions themselves often need to be made at a population level, with vaccination being the famous but not the only example. These are in tension with clinical Mainstream Medicine, as well as with much of the history of Western Medicine, during which most medical practice was private.

Nonetheless, given the very large amount of medical research and practice that Mainstream Medicine devotes to prevention, as well as the conceptual connection between stopping a disease before it happens and curing it once it has happened, I think it is not plausible to deny that prevention is a goal of medicine.

I suggested that it be seen as part of a single, overarching goal, alongside cure. What does that mean, and what is the argument for it?

The reason I see the goals of cure and prevention as part of a larger goal is that they are both in fact have the same effect if they are achieved—that is, the removal of disease. The only difference is when the intervention takes place. A preventive intervention takes place before disease; a cure, after its onset. This difference means that cure and prevention are not the same thing. But the *goals* of curing and preventing are subsidiary goals to the same end; if you have either goal, then you have the larger goal of favoring health over disease. We do not have a word for that, but if we did, I would use that, and say that were the goal of medicine. Since we don't, I say that the goal of medicine is the conjunction of the goal of cure and the goal of prevention. While cure and prevention themselves are different things, the goals of cure and prevention are part of the same, larger goal. That larger goal can exist prior to the onset of disease and can persist through its onset. It is the goal that many of us have, both when we are healthy and when we are sick, to not be sick; and it is the goal that caring parents have for their children. It motivates payment of medical insurance premiums as well as hospital bills. The idea that there are two different aims here is artificial. Prevention is what you do before it happens, cure is what you do after; but the goal of prevention is identical to the goal of cure: that is, bringing it about that disease is absent and health present.

This, at any rate, is my stance on the matter. It does not matter greatly for the arguments and discussions that follow if this argument is rejected, and prevention is seen as a separate goal of medicine. The

Curative Thesis, and my preferred alternative, the Inquiry Thesis, will need to be modified accordingly. My criticisms of the Curative Thesis primarily concern its ability to achieve the goal of cure. They might be extended to prevention, or alternatively it might be felt that we are a good deal better at prevention than cure. It does not matter for my criticisms of the Curative Thesis, for which it is sufficient to focus on cure alone. Accordingly I focus on cure in this chapter and the next.

2.7 CURE IS THE GOAL OF MEDICINE

It may sound dramatic to assert that cure is *the* goal of medicine, or even that it is the *ultimate* goal, to which any other goal is ancillary or secondary. However, to summarize:

- The pursuit of medical knowledge is in service of the goal of cure, not for its own sake;
- Pain relief is a use that society makes of medicine and a demand or expectation that it places on medicine, but not a goal of medicine in the way that cure is—one could have a medical tradition that did not aim to relieve pain but not one that did not aim to cure;
- The use of medical knowledge for the benefit of the patient is a culturally specific goal, since both the knowledge and the idea that we should benefit patients vary from culture to culture, even within the Mainstream tradition, and certainly outside it.

It is hard to think what other proper and legitimate goals, excluding cynical ones such as profit and power, medicine could have—other than prevention, of course, which I have argued are merely temporally different parts of the same larger goal of not being sick. I conclude that cure is the goal of medicine.

2.8 THE PUZZLE OF INEFFECTIVE MEDICINE

The Curative Thesis claims both that the goal of medicine is cure, and that curing is its core business. I have defended the first claim, but I find the second very hard to accept. In this section I will put forth a problem or puzzle; and in the next, I will give a more formal argument to the effect that the Curative Thesis must be rejected.

Imagine if blacksmiths were almost entirely unable to make things out of metal, haphazardly throwing chunks of metal into the fire, yanking them out again, and hammering them at random. Sometimes you might get something a little like the horseshoe you had hoped for, but mostly, you would just get a useless, misshapen lump. Yet imagine that blacksmiths had existed for thousands of years, like the medical profession: sometimes satirized, but often revered and accorded high social status; and that throughout all this time blacksmiths were routinely consulted by people hoping for horseshoes, pokers, and so forth. Imagine that kings and queens had royal blacksmiths, that there were rival schools, even that there were charlatans and quacks, despised by reputable blacksmiths. Would not all of this call for some explanation? How could there be quacks, if real blacksmiths were largely impotent when faced with a chunk of metal? What would a king or queen get out of having a royal black-smith if that blacksmith's productions were the result of largely sto-chastic hammering?

This, I suggest, is roughly the position of medicine. It has been largely ineffective in most times and in most places at delivering the cures that the people consulting it obviously want. And yet it has persisted, not merely as a ragbag of forlorn efforts, but as a profession, prominent enough to attract both great esteem and satire. How can this be? What have doctors been *doing*, if they have not been curing? How can they have been recognized as experts of any kind, if they have so often failed to deliver what those consulting them seek?

Now imagine that you get into a taxicab in New York City, and ask the driver to take you to the Empire State Building. Imagine the driver replied that she couldn't take you there, and you would have to walk. She recommends some shoes that would make the journey a little more comfortable on your feet, and tells you where you can purchase them (at your own expense of course). If the cab-driving profession was like the medical profession, then you might be a little disappointed, but all the same, you would thank the driver for her patient and sympathetic explanation, pay her a healthy sum, then get out and start walking, hoping you had enough left for the shoes.

This is analogous to your experience if you visit the doctor in a contemporary clinic with any one of a wide variety of common and unpleasant ailments, including the common cold (or upper respiratory tract infection), a sore back, or the blues. More serious ailments, notably cancers, will entail more serious treatments—not sneakers so much as a high-spec racing bicycle—but a cure will still often not be forthcoming. There remain many things that no contemporary medicine can cure. This is unsurprising. What is surprising is that we recognize experts in these areas nonetheless. We recognize oncologists, orthopedic surgeons, and psychiatrists, even though none of these has anything like a cure for every ill in their field. We recognize them as medical experts even regarding conditions they cannot cure. So what is their expertise? What is their competence—what are they good at, if not curing?

This is what I call the Puzzle of Ineffective Medicine: *given that medicine is extremely unreliable at achieving its fundamental goal, namely healing the sick, why does it persist?*

The Puzzle can also be framed for prevention, which is subject to parallel considerations. Smallpox inoculation was an early success, and contemporary vaccination is a strikingly successful preventive measure against some infectious diseases, just as antibiotics are effective cures for some infectious diseases. However, there are many other diseases for which we have no failsafe prevention recipe. Perhaps 1% of non-smokers will get lung cancer. Healthy, athletic, relatively young persons occasionally suffer coronary heart disease.

By way of example, consider what we know about nutrition. Obviously, we must eat and drink in order to live. But how does diet affect health?

Despite strenuous efforts to answer this question, our knowledge of nutrition remains extremely sketchy. Nutritional epidemiology is notorious among epidemiologists for being "garbage" (as I have heard it called), because residual confounding plagues almost every study. And reasoning from the theoretical effect that some substance is found to have in vitro has sometimes appeared worse than useless, for example in the some observations of an apparent *increase* in all-cause mortality risk among consumers of vitamin supplements.

Nutritional advice is notorious for contradicting other nutritional advice, or previous advice from the same source. Butter was bad for me until I was in my 30s, at which point, it was apparently good for me. Eating vegetables is often regarded as obviously good for you, but studies of vegetable eating are incredibly hard to control for socioeconomic class and geographical location. Moreover, anyone who has compared a "basic" tomato (one that is not part of a premium range) from an English supermarket with its counterpart in any Mediterranean country will appreciate that the vegetables themselves vary considerably from place to place. (Tomatoes in England are usually small, round, pale, hard, and tasteless, and they are typically imported. Tomatoes in Spain and Italy, for example, are usually large, lumpy-shaped, bright red, juicy, and delicious, and they are usually not imported. They have very different effects on the taste buds; so why not on health?) Red meat is virtually toxic according to some, while others maintain that meat of some form should form part of virtually every meal, and that the color really does not matter, so long as it is in its original form and not shaped like a sausage or a slice of bacon. (And again, any traveler knows that meat varies greatly from place to place; again, English meat is lackluster compared to South African.) Others say that even sausages and bacon are fine, but that legumes and sunflower oil, for instance, are dangerous. The truth of the matter is that, when it comes to nutrition, compelling evidence for or against the effect of some particular food or food group on longevity, health, or some specific characteristic (e.g., adiposity; muscularity) is hard to come by.

This is not to say that it doesn't matter what you eat; just that we don't really know much about how diet affects health. Not all nutritional advice is bullshit. But a lot of it is.

Nutritional ignorance is an example of a wider uncertainty. We do not have very much real preventive knowledge. Hygiene, vaccination, reasonable activity levels, and a varied, moderate diet are all obviously somewhat effective as ways to reduce the chances of a range of diseases. Some of these are ancient prescriptions, while others have been developed or innovated by Mainstream Medicine. Mainstream Medicine has improved our preventive abilities, but even on the most Whiggish view, it has hardly given us bulletproof vests. Moreover, these preventive platitudes are not easy to attain. War and poverty afflict the unfortunate; alcohol and psychological disorders mire the fortunate. Fortunate or unfortunate, almost nobody gets enough sleep, and almost everybody who is not starving eats too much, including many very poor people. We remain unable to prevent a great deal of disease, either because we don't know how, or because the only means of prevention we know about are not achieved. This is obvious from the fact that we still have so much medicine whose goal is to cure. Thus there is no solution to the Puzzle by saying that medicine has been successful at prevention where it has failed at cure; in fact, if medicine had been successful at prevention, the need for medicine whose goal was to cure would reduce. This has happened for some diseases; contemporary doctors are not experienced at treating smallpox, because they don't need to be. But for many other diseases, the goal of removing disease motivates continued attempts to cure.

The Puzzle can be framed for particular ineffective traditions past or present, and the answer might differ somewhat. Some medical traditions might exist or have existed through nothing other than fashion, foible, the folly of the rich, or the fiat of a tyrant. Nonetheless, no collection of piecemeal explanations can explain the general existence of medicine in all settled societies, and its persistence despite its widespread curative inefficacy.

Before going further, however, it is useful to formulate a more precise and emphatic argument against the claim that cure is the core business of medicine.

2.9 THE ARGUMENT FROM THE PERSISTENCE OF INEFFECTIVE MEDICINE

It would be ridiculous to pay a taxi driver for taking you nowhere, because the core business of a taxi driver is to take you to the place you want to go, typically in exchange for some payment. By applying a certain core competence (which may have multiple aspects or parts), which is something like navigating a car in the shortest time that is safely and legally possible between two points within a certain area, the taxi driver makes (or supplements) a living. They can make a living out of driving people to specified destinations because they can *do* it. It is what we expect of taxicabs, and we would not pay them if they didn't do it. But then it seems that being healed cannot be what we expect of doctors, because we often pay them regardless of whether they cure us. One cannot claim that something is one's core business if one does not reliably do it, and if one gets paid regardless of whether one does it. You can't make a living out of doing something that you can't do, to put the contradiction at its bluntest.

Where the business takes place under the auspices of a recognized profession, as has often been the case for medicine, the core business of the members of the profession typically involves the exercise of some skill, ability, or competence that the profession generally possesses, but the laity and other professions generally do not. So healing the sick cannot be the core business of the medical profession, even if it is the goal: again, to put the contradiction bluntly, you can't have a professional competence that is something you can't do.

Here is an explicit statement of the argument.

1. If the core business of medicine were to cure the sick, then medical traditions, disciplines, practices, interventions, or practitioners that were unable to reliably cure the sick would not persist. [The No Bullshit Premise]
2. But they do persist. [The Empirical Premise]
3. Therefore, it is not the case that the core business of medicine is to heal the sick. [Conclusion]

The first premise I dub the "No Bullshit" premise because it effectively asserts that medicine is not bullshit: that is, it is not a concoction of trickery, hope, and error. The corresponding objection to this premise is the Bullshit Objection, which, in short, is that medicine is mostly bullshit, and hence that its persistence shows nothing more than some combination of deceit, gullibility, and stupidity. This objection in effect argues that medicine could persist despite being unable to do its core business; or, alternatively, it concedes that the core business is not cure, but suggests that it is something utterly unrelated to cure, such as trickery. Although the latter version is not strictly an objection to this premise, it renders the conclusion argument a Pyrrhic victory. It is an effective objection to my Inquiry Thesis below, which seeks to identify some more benign competence that medicine might possess in the absence of cure.

The second premise is an empirical claim about the existence of ineffective medicine, and thus I call it the Empirical Premise. The objection to this premise is the Whig's Objection, which asserts that medicine in the past was unsuccessful, but contemporary medicine, in one of its forms, is in fact successful. The Bullshit Objection and the Whig's Objection can be combined, to argue that the core business is cure in one tradition (usually contemporary Mainstream Medicine) but bullshit in all other times and places.

From a technical perspective the argument is valid on the most common semantics for counterfactual or subjunctive conditionals (Lewis 1973a; Lewis 1973b; Stalnaker 1981; cf. Nozick 1981) and in any case the material conditional could reasonably be substituted. So if the No Bullshit Premise and the Empirical Premise are true, the Conclusion follows, and it is not the case that the core business of medicine is to heal the sick.

Before we get going, four qualifications concerning the notion of core medical competence.

First, medical competence may be distributed; I'm not concerned with working out whether each medical professional needs certain abilities or whether they may be spread about among different kinds of medical professionals. The latter scenario certainly seems to be the prevalent one in contemporary Mainstream Medicine, but for convenience I will often speak of "the medical professional" or "the doctor" and the competences that person must have, even though in practice I acknowledge that the competences could be distributed between different professionals and may not all be present in any of them.

Second, the competence of medicine could differ from the role it plays in society, or is expected to play. This is most obvious if medicine is all quackery, since then the role it is expected to play is to cure the sick, but its core competence is conning them.

Third, the explanation for the existence of medicine and for its continued utilization by sick people might be different from the immediate motivation that sick people have in consulting medical professionals. To repeat, I do not deny that sick people want to be cured. I am interested in understanding the existence of medicine given that cure is what people want, yet often isn't what they get.

Finally, I acknowledge that the core medical competence might differ in different times and places, and that my quest may therefore be hopeless. But in my view, it is always worth starting out hopeful. If we assume that the various medical traditions in different times and places have something in common that identifies them as medical, and then set about finding it, we may find something that we would not find if we started out pluralists. Implausible assumptions are sometimes the most productive.

2.10 CONCLUSION

Cure is the goal of medicine, yet not its core business. This leaves us needing to know what its core business is. In the next chapter, I argue that the core business of medicine is *inquiry*—that is, understanding and predicting health and disease.

Chapter 3

The Business of Medicine

3.1 THE INQUIRY THESIS

Lewis Thomas was a distinguished American physician who was trained in the 1930s, and thus worked through the curative revolution of the mid-20th century.[1] Describing his training, he wrote:

> We were provided with a thin, pocket-sized book called *Useful Drugs*, one hundred pages or so, and we carried this around in our white coats when we entered the teaching wards and clinics in the third year, but I cannot recall any of our instructors ever referring to this volume. Nor do I remember much talk about treating disease at any time in the four years of medical school except by the surgeons ... The medicine we were trained to practice was, essentially, Osler's medicine. Our task for the future was to be diagnoses and explanation. Explanation was the real business of medicine. What the ill patient and his family wanted most was to know the name of the illness and then, if possible,

---

1. Like the previous chapter, this chapter draws heavily on my inaugural lecture at the University of Johannesburg, "Prediction and Medicine," delivered in 2016, as well as the replies by Chadwin Harris and Thaddeus Metz. It also draws upon the subsequent paper, responses, and author's reply published in the *Journal of Philosophy of Medicine* (Broadbent 2018a; Broadbent 2018b; Harris 2018; Metz 2018).

what had caused it, and finally, most important of all, how it was likely to turn out.

(Porter 1997, 681–82)

Thomas is not saying that medicine's *goal* is explanation or prediction, but rather, that providing explanations and predictions is its core business (the latter more core than the former in his view). This is the idea that I intend to develop in this section, since, if it were true, then it might explain the persistence of medicine despite curative ineffectiveness, by identifying another kind of competence whose exercise might constitute its core business.

The Inquiry Thesis says that medicine is an inquiry into the nature and causes of health and disease, for the purpose of cure and prevention (that is, for the purpose of removing disease in favor of health). The Inquiry Thesis thus agrees with the Curative Thesis that the goal of medicine is cure. I have given my arguments for this claim in the previous chapter. However, it gives more detail about the character of medicine. The Inquiry Thesis asserts that medicine has the character of an inquiry. It is an inquiry for a certain purpose, and its purpose distinguishes it from science (whose purpose I do not want say more about than that it is different from medicine's). This purpose, cure, would explain why medical inquiries so often concern individuals, and the focus on explanations of individual cases of disease, in contrast to science, which typically aims for generality. (I have argued that even epidemiology, which is the medical science concerned with the spread of disease in populations, does not concern itself with general truths but with particular explanations of population-level events [Broadbent 2015a; Broadbent 2015c; Broadbent, Vandenbroucke, and Pearce 2016].)

To assess the Inquiry Thesis, we must identify the core medical competence whose application is the core business of medicine, and we must see if these competences do better against the Persistence of Ineffective Medicine. My task, then, is to show that inquiry can be explained in a way that shows that medicine has been much better at this than it has been at cure.

My hypothesis is that the core competence of medicine is *understanding*, and that this is demonstrated empirically by *prediction*. I don't mind whether the core competence of medicine consists in a conjunction of both understanding and prediction, or alternatively whether we say that there are two core medical competences. I tend toward the view that they are best seen as part of the same competence because of the connections between them, which we will explore shortly. But for clarity let us distinguish two theses.

The *Understanding Thesis* says that *the ability to engage meaningfully with the project of attaining understanding of health and illness is essential to medicine*. (I deliberately leave open, for now, whether the understanding relates to health and illness in general, or in the particular patient: we will discuss this important question in the next two chapters.) The *Predictive Thesis* says that *the ability to make good predictions about health states, including conditional predictions, is essential to medicine*. It might be helpful to think of it as follows: the core *intellectual* competence is understanding, and the core *practical* competence is prediction; alternatively, that the intellectual manifestation of the core medical competence is understanding, and the practical manifestation of that same competence is prediction. However, this distinction is not essential: for my argument, I do not want or need to put any weight on the distinction between intellectual and practical competence; and nor do I think that the distinction supplies any great insight into medicine, except perhaps to clarify how prediction and understanding might be thought of as part of one and the same medical competence.

I have two arguments for these theses. The first is by inference to the best explanation. The second is that, to have any plausibility, the Curative Thesis in fact requires the Understanding Thesis to be true, while the reverse is not so; and this asymmetry allows us to conclude that understanding and prediction are fundamental competences in a way that cure is not. I will explore the relation of prediction to both understanding and cure in the second argument. Meanwhile, please note that I do not mean to suggest that prediction is either necessary or sufficient for understanding.

3.2 AN INFERENCE TO THE BEST EXPLANATION

The first of my two arguments for these theses is by inference to the best explanation. This is a common inference form in everyday life, in many times and places, and it is also a common inference form in science, in which context it received its canonical theorization from Peter Lipton (2004). The idea is very simple: an inference to the best explanation occurs when we infer the truth of a claim on the basis that it would be the best explanation for something else. For example, I touch the kettle in my kitchen and find it is hot. I infer that my wife must recently have made a cup of tea. There is nobody else in the house who could boil the kettle; my children are too small. The characteristic feature of this inference form is that it is not failsafe. It is not a *deductively valid* inference. There are other possible explanations for the warm kettle. A cat might have tapped the button as it walked past. A stealthy intruder might have paused for refreshment. A peculiar electrical fault might have caused the kettle to boil itself. A ghost might have possessed the kettle. These are all conceivable and, in a logical sense, possible. But they are worse explanations for the warm kettle, for different reasons.

The project of understanding and justifying inference to the best explanation consists largely in spelling out what makes an explanation good, better, and best. This is not the project of this book. Provided that we agree on an ordering for the explanations that are available to us, we can use this form of inference without more theoretical detail than I have just provided.

I will present my inference to the best explanation first in relation to an imaginary, idealized example, and then in relation to an historical example.

Suppose you have a sore finger and go to see a doctor. The doctor recognizes the disease and gives you a detailed explanation of what is going on. She regrets that she can do nothing, but tells you that in three days it will turn green, and then fall off two days after that. She may be able to prescribe some painkillers, and offer some advice about how best to manage the situation; but she cannot offer a cure.

Here is a case where medicine has no cure. Nonetheless, it could still be a case of medicine being practiced. Indeed, it is medicine that *tells* you there is no cure. You do not doubt the competence of the medical profession as a whole, or the particular persons involved in this case, *merely* because the finger cannot be saved. Of course, you might be disappointed and seek second and third opinions, to make sure that your doctor is competent. But assuming that they, and the internet, all agree, you can accept that this is a competent doctor, with a competent medical opinion about this particular ailment, perhaps even an expert about this particular class of ailments (maybe even mentioned in a Wikipedia article)—who, nonetheless, is unable to save the finger.

What makes this a competent medical opinion? My proposal is that it is the fact that the doctor *understands what is going on*. She can *explain* what is happening to the finger. Perhaps she can also explain why she cannot cure it: one can explain why one cannot do something, as when a physicist explains why she cannot build a rocket that travels at light speed. In a case like this, your health may be little or no better off than before you sought medical help. It is not clear that you, as patient, will necessarily even obtain any understanding yourself;

you may be told some story, but it may be grossly simplified, or you may be unable to understand it. You may benefit from knowing what is going to happen, but then again, you may be wracked with worry and not benefit at all.

Indeed, you may get nothing you want or value out of the consultation. However, my interest is not in what the patient gets out of it. My interest is in what best *explains* the fact that you as patient, and right-thinking people at large, may still accept the doctor as (a) competent in general, (b) part of a legitimate medical discipline, profession, or tradition, and (c) having acted competently in this case. The explanation cannot be that the doctor offers a cure. Nor is it obvious that someone offering a cure will be deemed more competent than someone who is not offering a cure, until the effectiveness of that cure is confirmed by someone who is medically respectable, which further shows a distance between curative ability and medical competence. A person purporting to offer a cure might be dismissed as a quack. The recommended treatment may be supportive only.

My suggestion is that the doctor convinces us of her competence through showing that she is meaningfully engaged in an inquiry into the nature of the patient's disease and the reasons for it.

This may mean providing a diagnosis and prognosis. I will deal in the next section with the question of how "correct understanding" is identified, but in this case, accurate prognosis is one important piece of evidence, because of the fact that it at least *appears* to be directly available to the non-expert. Accurate prediction can be incredibly impressive, even in the absence of cure.

It is important to note that meaningful engagement with inquiry into the nature of the patient's disease and the reasons for it is also compatible with confessing ignorance and puzzlement. Even in cases of puzzlement, however, the competent doctor will be able to *exclude* a number of conditions that she does understand. Her ignorance itself is interesting, even informative, because she knows so much. To be *meaningfully engaged* with the inquiry into the patient's disease implies two things. It means, first, having (or having access to) a substantial body of medical knowledge, and second, being able to apply it to the particular patient in question, to work out where in that large body of knowledge the particular case fits. The doctor does not need to succeed at fitting the patient into the body of medical knowledge to be meaningfully engaged, but she does need credible expertise in that body of knowledge, as well as credible diagnostic skills—credible enough for a failure to fit the patient into the body of knowledge is informative in its own right.

Now for a real historical example, where a dramatic and painful diagnostic intervention was undertaken, yet there was little therapeutic benefit of doing so. This example was given to me by Jan Vandenbroucke, one of the leading lights of clinical epidemiology in the past few decades. Although his status and background are relevant to the *interest* of what he has to say, I do not take him as an *authority* in this context, but rather as an actor in the medical world, as a sociologist might. I put some of my ideas to him in private communication, and he responded informally, in private communication, as follows.

Vandenbroucke's parents were both doctors in Belgium. His father began practicing before World War II, and his mother's father practiced much earlier. Both were trained and started practicing when there was no penicillin, no dialysis, no antihypertensives, no drugs for tuberculosis—indeed, very little at all by way of cure. There was some surgery but it was haphazard, and generally on the basis an "educated guess" given the lack of preoperative scanning technologies. Nonetheless, insists Vandenbroucke, his parents, and grandparents, were occupied full time with making diagnoses and predictions about the course of disease.

Vandenbroucke gives the example of a child with meningitis. Such a child would receive a lumbar puncture:

> if the fluid was clear, they could tell the parents that it was probably viral, and that the child would most likely be OK in a few days; if the fluid was opaque, they determined whether there were bacteria, in which case the child had a slim chance with sulfonamides, or whether it was tuberculosis meningitis, in which case the child would die a slow and painful death— those latter children were isolated and perhaps sedated with barbiturates (the only type of sedation they had, next to opiates). So actually, the lumbar puncture and the ensuing diagnosis had almost no consequences for therapy—but they were done to understand, to find an explanation and to be able to tell a prognosis.
>
> (Vandenbroucke 2016)

Vandenbroucke's example suggests that decisions of some sort were taken on the basis of the diagnosis obtained by lumbar puncture, and attempts at intervention were made. They ranged from the feeble to the totally ineffective, and some were not interventions but decisions about the best course of action (e.g., isolating certain children).

One might insist that such interventions were nonetheless cures, by my lights. Such is the nature of real-life examples; they never quite fit your hypothesis, which is why philosophers and scientists alike gravitate toward idealizations. Nonetheless, I think this example cannot be fully explained unless the Understanding Thesis is correct. The overriding point is that the effectiveness of these therapeutic interventions really does not seem to warrant or fully explain the lengths gone to in order to obtain a diagnosis. A lumbar puncture is a big deal, and one could perhaps manage the minimal treatment options described above without the positive diagnosis, by giving sulfonamides to all on a precautionary basis, and if children started to suffer severely, isolating them and giving them barbiturates. Moreover, Vandenbroucke asserts that he could "tell you countless stories like this," and I will suggest three further examples from contemporary Mainstream Medicine below.

I submit that these examples show that there can be and often are diagnoses and predictions made without significant curative import, and moreover, these are often taken as displays of medical competence. Clearly there is a need for more thorough empirical (sociological, historical) inquiry. However, I submit that the best explanation of the examples discussed here is that meaningful engagement in the project of understanding health and illness and the reasons for them is the core medical competence, and thus that these examples are evidence for the Understanding Thesis.

3.3 THE PRIMACY OF UNDERSTANDING AND PREDICTION

My second argument is to the effect that understanding and prediction are primary, and cure secondary. Prediction is *logically* prior to cure, because you cannot have cure (I shall argue) without prediction, whereas you can have prediction without cure. Understanding is *temporally* prior to cure, because as a matter of historical fact we have had considerably more understanding than we have had curative ability.

Let me now spell the argument out in more detail.

Suppose a sick burglar decides to pick a few drugs at random from a pharmacy he is burgling. As it happens, he gets something that works. Has he cured himself? Perhaps, but he cannot claim credit for curing himself. To claim the credit for curing somebody, you must exclude the possibility that you got lucky. At least, if you want to claim that you possess a special competence that is curing people, and that you cured the person through the exercise of this competence, then you must exclude cases of luck. This is because, by definition, getting lucky is not a competence; it is luck.

If lucky cure is ruled out, which it must be for the Curative Thesis to be plausible in the first place, then the Curative Thesis entails the Predictive Thesis.

In order to accept that a certain contemplated future intervention will cure, one needs to accept two predictions: first, a prediction of recovery conditional on the intervention; and second, a prediction of doing worse conditional on not taking the cure. After taking the cure, one will only accept that a past intervention has cured if one in fact recovers as predicted, and second, if one believes the doctor's claim that had one not taken the cure, one would not have recovered. Thus cure can only be a core medical competence if prediction is also a core medical competence, since cure cannot be present without good prediction: even if an intervention is effective, it is not a cure if its effectiveness was unpredicted. Cure, in the only sense that would make the Curative Thesis plausible, must allow us to confer medical competence, and so cannot be just getting lucky. This is why prediction is logically prior to cure (meaning non-lucky cure of the sort associated with the Curative Thesis).

Understanding is *temporally* prior to cure, as a matter of historical and contemporary fact. Understanding can be, has been, and often still is present without curative ability. Porter writes:

> Biomedical understanding long outstripped breakthroughs in curative medicine, and the retreat of the great lethal diseases (diphtheria, typhoid, tuberculosis, and so forth) was due, in the first instance, more to urban improvements, superior nutrition and public health than to curative medicine. The one early striking instance of the conquest of disease—the introduction of

the first smallpox vaccination—came not through "science" but through embracing popular medical folklore.

(Porter 1997, 11)

Western Medicine accumulated enormous quantities of knowledge over a period of centuries, with almost no improvement in curative abilities. Well into the early 1900s, there was almost nothing that a doctor could do about almost any disease (Porter 2002, 37). Some improvement had been achieved by then (hands were being washed, blood was not being let) and there were a handful of moderately effective interventions: "quinine for malaria, opium as an analgesic, colchicum for gout, digitalis to stimulate the heart, amyl nitrate to dilate the arteries in angina and, introduced in 1896, the versatile aspirin. Iron was ladled out as a tonic, as were senna and other herbal preparations as purgatives. True cures remained elusive, however" (Porter 2002, 37).

The real medical assistance (as opposed to "eyewash" prescriptions) provided by the medical profession remained minimal even in the early 20th century: advising relatives to let the patient rest, to try to get the patient to drink a little, to mop the brow until the fever passed, and so forth. Most of the magic bullet cures that we associate with Mainstream Medicine arrived after the first decade of the 20th century. By this point we already understood a very great deal, and had done so for a long time.

Predictive ability likewise outstripped curative ability considerably. In the Hippocratic treatise *On Prognosis*, we find this remark:

If [the physician] is able to tell his patients when he visits them not only about their past and present symptoms, but also tell them what is going to happen, as well as fill in the details they have omitted, he will increase his reputation as a medical

practitioner and people will have no qualms about putting themselves under his care.

(Lloyd 1983, 170)

In similar vein, Vandenbroucke recalls:

> Doctors gained fame, and were awed, by the fact that they had
> predicted rightly (even if the prediction was some very bad out-
> come). Some gained posthumous fame by predicting their own
> diagnosis and course of disease and asking their colleagues to
> confirm at autopsy (like Trousseau who predicted his gastric/
> pancreatic carcinoma . . .).

(Vandenbroucke 2016)

One might wonder how (and doubt whether) predictions could be
accurate if the understanding they were based on is, by modern lights,
mistaken; and I will deal with this question in section 3.7.

A further reason to think that prediction and understanding are
primary, and cure is secondary, is that the former two stand in a symbiotic relationship to each other, but not to cure. Prediction is neither
a necessary nor a sufficient condition for understanding. However,
if you can demonstrate some predictive ability, you have *favorable
evidence* that you understand; and if you altogether lack predictive
ability, this is *not favorable* toward the claim that you understand.

The reverse holds true too: predictions are more credible if they
are accompanied by a convincing claim to understanding. This is true
both beforehand, when trying to explain someone of the truth of your
prediction, and afterward, in order to convince someone that your
prediction was a good prediction and not a fluke. One of the reasons
that economists who claim to have predicted this or that financial
event are so rarely credible is that there are so many economists,
making so many different predictions, that one is inclined to suspect

that the real reason for the accurate prediction was just that this economist was searching for something to say that others had not already said. This suspicion will be somewhat allayed if the economist can "show their working" and convince us that they really knew what was going on, and that their prediction stemmed from this. Of course, a good track record will also help credibility. But this is because the ability to repeatedly make good predictions is an indicator of understanding. If we did not think there were understanding, we might be inclined to appeal to the same stochastic explanation. (So, if 100 economists predicted the first event right, maybe 50 predicted the next one so badly wrong as to damage their credibility; and so on, until we are left with this one. I have a friend in a hedge fund who asserts that many careers are better explained in this Darwinian style than by any unusual predictive ability, and empirical research into predictive ability in relation to sociopolitical events seems to bear out the suggestion that many putative experts are bad at it, while the best ones are still not that good [Tetlock 2005].)

I do not say that these connections are invariable. The point I rely on in this argument is that we do not usually attribute predictive ability to someone without also either previously accepting or simultaneously attributing understanding. If someone takes a look under the hood of your car and tells you that in the next couple of hundred miles you will blow a gasket, and then you do, you are likely to suppose that the person knew what was going on. If someone makes a lot of money on the stock market, year in, year out, you are likely to suppose they have some insight into how markets work. And if someone looks at your finger and tells you it will fall off a week from now, and it does, you will be likely to suppose that they understood something about the finger. What is more, in each of these cases, if you attribute the predictive success to luck rather than skill, you will probably also deny that understanding was present. The two are not *necessarily* linked in every case. It could be bare experience, and enumeration of similar cases, that enables a prediction, and sometimes this will be the better explanation than the attribution of understanding or insight. However, very often, some degree of understanding is the most plausible explanation for a good prediction, especially in varying circumstances, or conditional predictions concerning the outcome given some contingency. In short, understanding and prediction are often taken as signs of each other.

The situation, then, is as follows. Predictive ability is necessary for a cure to count as non-lucky, and for this reason, the Curative Thesis entails the Predictive Thesis. Understanding and prediction stand in a symbiotic relationship, such that they are generally taken as signs of each other. Thus the Understanding Thesis is very likely true if the Curative Thesis is true. It would be strange to accept that a doctor is competent on the basis that she can cure, but to deny that she understands. The reverse entailments do not hold, however. Doctors may be able to understand and predict even without any curative ability. Moreover, the historical record shows that both understanding and predictive ability far outstripped curative ability.

This allows us to claim a kind of primacy or fundamentality for understanding and prediction in medicine. Where we attribute cure, we typically (often, not always) attribute understanding and predictive ability also. But not vice versa. The Curative Thesis says that cure is the core medical competence. The proponent must then also accept that understanding and prediction are also core medical competences. The proponent must then insist that, where these are present without cure, there is no medicine. Since there is so much medicine without cure, this is quite implausible. But having admitted that understanding and prediction are core medical competences, because of their relationship to cure, the proponent then needs a good reason not to accept that understanding and prediction remain core medical competences, given their presence in cases of medicine without cure.

The main reason to deny this would be to deny that the presence of understanding and cure in ineffective medicine: to argue, in short, that medicine has been just as bad at prediction and understanding as it has been at cure. We will consider these objections in relation to the Whig's Objection below in section 3.6.

3.4 THE BULLSHIT OBJECTION

The No Bullshit Premise says that, if the core business of medicine were cure, then medicine would not persist if it were ineffective at cure. This is plausible if we can assume that people have some sort of a handle on whether interventions are effective. The main objection to this premise is that this is not plausible, because a tradition or intervention might persist because it is *wrongly thought* to be effective.

This could come about in various ways. Doctors might be deluded along with their patients; or doctors might deliberately delude patients. However, I think the most plausible version of the objection is that doctors are simply *careless* about effectiveness, not really knowing whether their treatments are effective or not, but all the same telling their patients that they are effective.

Harry Frankfurt develops the notion of bullshit in all seriousness as filling a gap that he perceives between telling the truth and deliberately lying (Frankfurt 2005). A bullshitter is someone who speaks without regard for the truth, in order to achieve some end or other. The Bullshit Objection thus is not that doctors deliberately deceive patients, but merely that often, doctors don't know whether what they say is true, but say it all the same, for some end unconnected with truth—presumably financial in most cases. The Bullshit Objection is more plausible than asserting that medicine is a huge lie, because one can only lie if one believes that what one is saying is false. In many cases, this would be implausible for medics: a doctor might not know whether some course of action will cure, or not, in which case he cannot be positively lying when he asserts its efficacy (though he may lie about his confidence in it). "Bullshit" is a more appropriate term, since it is compatible with the doctor not knowing whether what he says is true, or even believing that it is true, but nonetheless speaking without regard for truth. David Wootton sums it up:

> For 2,400 years patients have believed their doctors were doing them good; for 2,300 years they were wrong.
>
> (Wootton 2006, 2)

This line of thought derives credibility from the phenomenon of "self-limited disease," a 19th-century term (Bynum 2008, Chapter 1), combined with the common fallacy of post hoc ergo propter hoc (after, therefore because of). The majority of diseases heal in time, many in just a few days. Combined, this empirical fact and this common fallacy in our thinking may enable many ineffective or even harmful medical interventions to persist for the reason that they are wrongly thought to be effective or beneficial.

My favorite example is the use of high-dose vitamin C to pre-vent or cure the common cold. If you experience the symptoms of a common cold, and then take vitamin C, you will probably find the cold goes away in a couple of days. However, you will probably have exactly the same experience if you do not take vitamin C. It is well established that high doses of vitamin C do not in fact assist with preventing or curing colds. The idea has absolutely no medical basis at all: clinical trials do not bear it out; no biomedical theory supports it; and it does not have its roots in any ancient tradition, folklore, or other body of non-Mainstream medical knowledge. The potency of vitamin C is, in fact, the brainchild of an individual enthusiast, Linus Pauling, who somehow managed to get a popular foothold for this particular idea. His claims for the benefits of mega-dosing, including in relation to the common cold, have been discredited (Hemilä 2009), despite remaining popular. The uptake of this idea was not because vitamin C is effective against colds, but because of the susceptibility of assessments of its efficacy against this self-limited disease to the fallacy of post hoc ergo propter hoc. Indeed, the only true cure for the common cold is a little-known remedy that I have personally perfected, and that I alone know about. Interested readers (who are prepared to pay a very reasonable fee) should not trouble the publisher with inquiries but should contact me directly.

There is thus no doubt that medicine is *sometimes* bullshit, and that this *sometimes* explains the persistence of ineffective medicine. Does this defeat the Argument from the Persistence of Ineffective Medicine, and is it a problem for the Inquiry Thesis?

3.5 NOT ALL INEFFECTIVE MEDICINE IS BULLSHIT

Even taking self-limited disease and the temptation of post hoc ergo propter hoc (after, therefore caused by) into account, I maintain that there are also enough cases where all parties are aware of the inability of medicine to cure for a further explanation to be required. Contrary to the Bullshit Objection, the ineffectiveness of medical interventions, disciplines, or traditions both was and is commonly known to both practitioners and patients. Yet they were and are commonly regarded as respectable and medically expert nonetheless. The situation is thus one where *both* bullshit *and* respectable medicine exist. My reply to the Bullshit Objection will be that it is the latter that explains the persistence of ineffective medicine, with bullshit riding as a passenger.

First, at least sometimes, the ineffectiveness of medicine in the past was known. Of classical Hippocratic doctors, Porter writes that they "made no pretence to miracle cures, but they did pledge above all to do no harm . . . and presented themselves as faithful friends to the sick" (Porter 2002, 30). Of the late 19th century, Porter writes that "doctors knew their prescriptions were largely eyewash" and further that "churchgoing folk did not expect the family doctor to perform miracles" (Porter 2002, 39). There is a lot in between antiquity and the Victorians, and I do not doubt for a moment that there have been many false hopes knowingly or carelessly raised. But there have *also* been at least *some* honest doctors, who were *also* aware of *at least some* of the limitations of their curative abilities.

Second, ineffective medicine is not only a thing of the past. It persists today. Wootton regards it as obvious that "modern medicine works," but this is vague, and, on natural resolutions of the vagueness, false. It is certainly not the case that all modern medicine is effective at curing some disease. It is equally certainly not the case that for all diseases, some modern medical cure exists. These are the obvious resolutions of the vagueness of the phrase, and they are obviously false.

For my purposes, I am not interested in deciding whether a given tradition as a whole works, or is effective, or anything of the sort. This is a question about how we judge the success of a tradition, and the Argument from the Persistence of Ineffective Medicine does not rely on making any general judgments of this kind. All it requires is that an appreciable quantity of curative ineffectiveness persists in every medical tradition. This is certainly the case for Mainstream Medicine, regardless of whether we feel inclined to assent to the general claim that Mainstream Medicine "works."

It is obvious that Mainstream Medicine is unable to cure the lethal diseases that top the charts for cause of death in developed countries. Cancer and heart disease are obviously diseases for which we have no cure. One might object that, when people get old, they must die of something, and that medicine cannot be deemed ineffective merely for failing to deliver immortality. But even if one excludes people above, say, 70, in developed countries, one finds disease claiming lives at every age. The numbers are fewer than they were in the past, but as I have indicated, I am not interested in establishing some general proposition to the effect that Mainstream Medicine doesn't work. I really don't care whether it works or not. The point is that it contains a large amount of curative ineffectiveness; enough to call for an explanation, given that its goal is cure.

Rather that focus on the obvious and commonly lethal diseases, I want to consider three examples of more mundane diseases, against which Mainstream Medicine is not strikingly more potent than Western Medicine of 100 or 1,000 years ago. These are all cases of which I have some direct experience, either relating to myself or people close to me, and I suspect that all my readers do, too. There is a reason to appeal to personal experience, which is that there is sometimes a gap between how people reason about medicine as an object of study and how they think or behave in medical matters relating to themselves (which is probably just as well).

First, doctors acknowledge that there is nothing they can do to cure the common cold. Some pain relief, and something to dry up a runny nose, perhaps; but there often remain symptoms for which no effective intervention is available. A cough often seems (in my experience) to be entirely untreatable.

Do not reply that the cold is not sufficiently serious for us to bother trying to find a cure. If we could cure the cold, or vaccinate against it, we would. It would be very profitable for the company that developed the cure. Moreover, the cold can claim lives among the infirm, elderly, young, asthmatic, and so forth. My daughter, who is probably asthmatic (as are many in my family), was hospitalized five weeks after birth with a chest infection that began as a cold. She choked several times on her own mucus and could have died. Her blood oxygen levels were very low when she was admitted. She was given oxygen and acetaminophen (paracetamol). When she choked, and occasionally when she just seemed bunged up, an electric vacuum pump was used to suck mucus out of her nose. You can also do this using your mouth, so this was a convenience rather than a necessity. It is probable that the baby would have survived without the acetaminophen, because her temperature was not excessive; but it made her much more comfortable. The oxygen surely reduced the chances of fatality, but again, the child would most likely have made it through without. Again, it would have been much less comfortable: we would have been constantly worrying about her oxygen levels, waking her up, and so forth. The temperature in the ward was carefully controlled, which reduced the overall stress on the child. This is supportive treatment, and very good supportive treatment at that. It is better than the supportive treatment of 100 years ago, and a lot better than that of 1,000 years ago. Yet, from the point of view of a *cure*, there is no improvement on olden times: there was nothing that significantly altered the course of the disease for the better. The infant was not cured; she did not even improve for the first couple of days of her five-day stay. But she was in an environment where it was much harder for her to die, and she was supported and cared for until she recovered. The common cold is not something we are able to cure, even in a good modern hospital, despite being an irritation to many and dangerous to a few.

Second, consider herniated ("slipped") lumbar disc. Painkillers are often ineffective for the "nerve pain" caused by a herniated disc, as anyone who has experienced it will report. It is possible to gain some temporary relief from an epidural. There are surgical options, such as fusing the vertebrae, or cutting off the protruding part of the disc. However, these are arguably pain-relief measures rather than cures; fused vertebrae cannot hinge, and cutting a protrusion off a disc still leaves the rest thinner than it would otherwise be, causing pain in some situations (e.g., impacts, bumps). For these reasons, the surgeon will typically—and ought to—acknowledge that surgery is a last resort. Moreover, she should also acknowledge that these interventions have a shelf life. They do not usually keep pain at bay forever, only for a few years. Surgery has its place, because sometimes pain is unbearable, or interferes with working life, and so forth. I am not *denigrating* surgery for herniated disc. I am merely stating what it actually achieves; and that is emphatically not cure. These interventions do not actually repair the herniated disc. They do not appreciably alter the course of the disease for the better, but rather alleviate the pain of the disease, for a period of time. Many non-surgical interventions are recommended for back pain, but none is particularly effective.

Third, consider depression. Suicide is among the top ten killers in many countries, as it was a hundred years ago (Rockett 1999): the idea that psychological ailments are a symptom of modern life is as mistaken as it is old. Although the nature of depression is complicated, and the causes are extremely hard to untangle, there is no doubt that it is a serious affliction. Again, anyone who has ever suffered a bout of serious depression—and that is a large number of people—will understand this.

Pharmaceutical treatments for depression exist, and they can be quite effective. There are those who deny this; for example, Jacob Stegenga argues that the evidence for the effectiveness of selective serotonin reuptake inhibitors (SSRIs) is too weak to warrant their prescription given the risks and side effects. This is not the dominant medical view. All the same, it is doubtful whether SSRIs are a cure for depression. One is not rid of the causes of depression after taking a course of antidepressants, as one is rid of bacteria after a course of antibiotics. They alleviate the suffering of the disease, or else they modify its symptoms, or perhaps like insulin for diabetics they supplement the body in a way that enables it to function properly. But it is not clear that these things amount to altering the course of the disease for the better. And the same can be said for other treatment options, which range from losing a bit of weight, taking more exercise, or taking it easy at work, to more heavy-duty drugs and institutionalization, including against your will.

What is striking about at least the latter two afflictions for which Mainstream Medicine lacks highly effective interventions (herniated disc and depression) is that they are not isolated. They are part of entire fields of specialization that are relatively ineffective. In relation to the herniated disc, we have orthopedic surgery. In relation to depression, we have psychiatry. In the case of the common cold, it is less clear that we have a specialized field—perhaps virology. However, what is clear is that the early successes of antibiotics against bacteria failed to be replicated against viruses, and that Mainstream Medicine remains very limited in what we can do about viral diseases. To this extent, the common cold is typical.

Let me summarize my response to the Bullshit Objection. I deny that ineffective medicine has persisted only because it is bullshit, even while admitting that bullshit is the explanation sometimes. I hold, first, that ineffective medicine in the past was recognized as ineffective, and yet also as medicine; and second, that ineffective medicine remains part of Mainstream Medicine, and that this is recognized by both patients and the profession. The ineffectiveness may not be *named*: people might not say, "This is not a very effective intervention/field." And it may be normalized: people might say, "But this is just what effective treatment of a slipped disc is!" Or they might blame the failure on the disease: "Depression is very difficult to treat." Nonetheless, in factual discussion of the likely outcome of a proposed course of treatment, I maintain that it is common enough for medical persons to be honest and undeluded, and for patients to likewise know what to expect from the treatment proposed. In some areas of Mainstream Medicine, we have learned not to expect miracles, just like Porter's churchgoing folk. Trickery, hope, and error are not constant companions of ineffective medicine, and thus cannot explain all of its persistence, even if they explain some; thus the Bullshit Objection fails.

3.6 THE WHIG'S OBJECTION

The Empirical Premise of the Argument from the Persistence of Ineffective Medicine can also be challenged. You could agree that ineffective medicine has persisted, and does persist in many places, yet object that not all medicine is ineffective, and that there is a clear difference between the effective and the ineffective kinds of medicine. You could then argue that the problems arising from the persistence of ineffective medicine do not afflict the effective kinds or parts of medicine. I call this the Whig's Objection, since in essence it asserts that the present is different from the past, and that we are now succeeding where we have failed before, and that we are now on the royal road of progress. Parts of the history of medicine are likely to be seen as earlier steps along that royal road; but most of it will be discarded.

This is exactly Wootton's view of contemporary Mainstream Medicine. Wootton argues that, for most of history, Western doctors have done more harm than good. He argues that this began to change in 1865 with William Lister's successful use of antiseptic procedures in a surgical procedure. He calls medicine prior to this time "bad medicine," characterized by doing little good and more harm. He is clear that he is referring to "bad medicine that was honestly believed to be good medicine," conceding that "There have always been incompetent, careless, and even malevolent doctors, but what I am concerned with in this book is the medical profession at its best" (Wootton 2006, 26).

Wootton's stance is Whiggish, and perhaps unashamedly so; to be Whiggish is not necessarily to be wrong, even though the term is often used as an intellectual insult. To be Whiggish is to apply the standards of the present to the past, and thus to see history as a gradual but inexorable path of progress toward truth as the veil of ignorance is lifted by the efforts of great minds. In being Whiggish, especially in applying the standards of the present to the past, Wootton's view is in tension with the avowedly anti-Whiggish stance of the historian of medicine, William Bynum:

> It seems to me that those in the past who had access have generally sought medical care that was on offer, and believed that there were good doctors and bad doctors. They wanted a good doctor to take care of them. So do we. What has changed is what constitutes a "good" doctor.
>
> (Bynum 2008, 328/2580)

Of course, there is a balance to be struck here. If there is a spectrum along which historians apply the notions of success and progress as if they were timeless and objective, then Wootton is at one extreme. At the other end of the spectrum lies an equally extreme kind of relativism. Neither is appealing, the former because it fails to take account of Bynum's point that success is relative to a goal and the shared goals of the past may differ from ours, and the latter because it removes the very strong sense we have that certain experiences are universal and certain facts objective. Experiences of pain, suffering, and disease are probably among these. Yet medicine is clearly shot through with cultural influences.

The central challenge for the second part of this book is reconciling the apparent universality of experiences of health and illness with the evident locality of medicine and its concepts. For present purposes, the point is simply that if you incline to the Whiggish view, you might dispute the Empirical Premise, and seek to demarcate an effective tradition or branch of medicine, to which the Argument from the

Persistence of Ineffective Medicine does not apply. For purposes of argument I will assume that the identified branch is contemporary Mainstream Medicine. The strategy of my response can be applied to other kinds of medicine too.

My response to this form of the Whig's Objection has largely been covered already in the response to the Bullshit Objection. Contemporary Mainstream Medicine contains much ineffective medicine. To reiterate, I have no interest in deciding whether this or that tradition as a whole "works" or "is effective." I think that is a blind alley (one we will discuss in connection with Medical Nihilism in Chapter 6). My response, rather, is that there are recognized disciplines within Mainstream Medicine with little curative efficacy to boast of, and recognized experts in these disciplines whose medical expertise is not doubted merely because there is little or nothing that their specialism enables them to cure. The impotent medical expert is alive and well in Mainstream Medicine. Nor does the impotent medical expert thrive only in "bad medicine": a psychiatrist is the best person to go to if you are seriously depressed, not because they can cure you, but because they can direct your care. This was true for good Hippocratic doctors, and it is true for a good part of medical practice today.

3.7 THE WHIG'S OTHER OBJECTION

Hearing my response to her objection, the Whig might put a different but related objection, with the Inquiry Thesis as its target. The Whig's Other Objection says that past medicine was no better at understanding or predicting than it was at curing. Just as I have argued in response to the original Whig's Objection, understanding was built up over a long period of time. But go back a thousand years, and we have diseases being explained in terms of an imbalance of humors. Humors don't exist, and don't explain health or disease. Explanatory success despite curative failure thus cannot be the reason for the persistence of ineffective medicine. In addition, predictive success tracks explanatory success and there is no reason to believe that past predictive success was any better than past curative success.

Thaddeus Metz has objected to the Understanding Thesis along these lines.

> it is only recently that Western medicine has truly grasped some of the chemical, biological, and psychological facts that ground disease. And yet there has been Western medicine for much longer. By the logic of Broadbent's argument against the curative thesis, then, understanding is likewise not a core competence of medicine.
>
> (Metz 2018, 278–9)

For example, explanations in terms of humoral balance are not real explanations because the humoral balance theory of health and disease is false, and there are no humors.

There are two kinds of case to consider. One is the case where I want to insist there *was* understanding, but it did not yield cure. The other is the case where I agree that there was *not* understanding by modern lights, and yet I want to maintain that the Understanding Thesis is true. I will consider each in turn.

As is obvious from what I have said so far in this chapter and the last, I maintain that understanding outstripped curative ability in the past, and indeed that it still does so. It should be clear, then, that I do not believe there is a linear relationship between progress in understanding and cure. The medical revolution was a threshold phenomenon, occurring not because of growth of understanding per se, but because the pieces of the puzzle were all finally in place.

In the case of surgery, for instance, effective pain relief, sound anatomical knowledge, and effective antiseptic measures are all necessary. These arrived at different times, and via different routes. Growth of understanding in anatomical knowledge alone yielded no real improvement in the curative powers of surgery while it remained a horror worse than the affliction itself, since the surgeon had to work extremely fast, and the patient had to be extremely brave or desperate. Even when somewhat effective pain control became available, surgery remained ineffective because patients commonly died of septicemia. It was only when infection was also sufficiently well understood (following Lister's demonstration in Glasgow in 1865) for effective antiseptic measures to be devised that the surgical profession began its climb from the barbarism of the surgeon-barber toward the pinnacle of the medical profession that it occupies today.

Thus while I agree that understanding begets cure, I think that it does so in a stepwise fashion, and not linearly. This picture helps to make it more plausible that the explanations of the past could have been true, as I maintain they often were, and yet curatively useless.

My first response, then, is to emphasize that understanding can in principle exist and has in fact existed without curative ability. This first response does not, however, help me in cases where past medical explanations were clearly false by present lights. Thus I make a second response as follows.

I accept that the examples to which Metz gestures, including the humoral theory of disease and health, amount to explanations in terms of falsehoods. I also accept the general principle that an explanation in terms of falsehoods is not a real explanation, and does not offer real understanding. (Philosophers of science usually consider truth a necessary condition of explanation, because it is hard to see how a falsehood could explain anything, at least in a literal sense; perhaps there are metaphorical or literary or symbolic explanations, but these are not the kind that science seeks.) However, I do not think this means that understanding and cure are on a par in the history of medicine. In particular, I think that explanatory failures have a different significance to curative failures.

My response is parallel to a certain line of response to the objection to scientific realism known as the pessimistic meta-induction from past falsity (Laudan 1981). This is a fancy name[2] for the simple idea that, because past science was false, it is not reasonable to believe that present science is true, and in fact reasonable to believe that it is probably false. The argument challenges our idea that current science represents the accumulation of years of inquiry; it says that, actually, scientific inquiry has slumped disgracefully from one error to another. One line of response to this argument is to argue that past science is in fact more continuous with present science than the argument makes it sound. On this line of response, science *is* cumulative after all. Various forms of semi-realism, such as structural realism (Worrall 1989), have been proposed to try to make sense of what it is that science can retain and accumulate through successive, dramatic theory change. On this line of response, the past falsity of science does not amount to a complete failure.

It is this idea, rather than the idea of identifying some true content (structure, or something else), that I want to take from this debate, and apply to medicine. What I take is the idea that an intellectual project can progress, in some sense, even while a strict assessment of its assertions at a particular time would have to conclude that they were largely or completely false.

In the medical case this is perhaps even easier to see than in the scientific case. At least some branches of medical science lend themselves to being seen as cumulative over a fairly long period. In particular, anatomical knowledge is naturally understood as cumulative.

---

2. The paper in which the idea is presented also includes the word "confutation" in the title, for reasons that confute me.

Ancient views were very different from ours, but they were about the same subject matter—the body—and investigation eventually refined and corrected them.

However, I want to press a more general point, which is that cure and understanding are quite different, just as intellectual inquiry is quite different from practical action. Descartes' attempt to prove the existence of God was fallacious. But that does not make him a failed philosopher. Newton was a great physicist despite being mistaken about everything he theorized about. Copernicus was a great astronomer despite believing that the sun was the center of the universe. Einstein is not reckoned a bad scientist for refusing to accept that God plays dice with the universe. And so forth. If, however, any of these luminaries had designed a bridge that had collapsed under its own weight, that would probably make them a failed engineer (assuming the objective was to cross a gap and not to test a concept, construct a trap, etc.). Conversely, an explanatory effort can fail to provide true explanations, without the explanatory enterprise as a whole failing. Sometimes we even call that progress.

It is in response to this objection of Metz's (to Broadbent 2018b) that I have formulated my Understanding Thesis to refer to engaging with the *project* of understanding health and disease. I maintain that this project can exist even while it is not very good at providing truth, because, in general, the success of an explanatory project is not measured by its ability to provide truth.

Turning now to prediction, it is a familiar point in the history and philosophy of science that predictive success *is* compatible with explanatory falsity. Newtonian mechanics is predictively successful despite its strict falsity. Likewise, a medic in any tradition with a sound body of experience, and a training based on the experiences of others, may develop quite reliable predictive abilities—better than those of the average person. These, I suggest, are enough to keep him in business. His explanations for his predictive success might be quite false, but empirical facts massively underdetermine theoretical claims, and in any case, the real driver for the predictive ability would probably be experience rather than derivation from any sort of theoretical knowledge. This may well be true in modern medical practice too, on many occasions: the emphasis laid in medical training on practical experience suggests a strong continued role for tacit knowledge.

3.8 CONCLUSION

The Argument from the Persistence of Ineffective Medicine shows that the core business of medicine cannot be cure. If we accept that the goal of medicine is cure, as I do, we must come up with some other competence that the medical profession might possess. That competence, I suggest, has two parts. One is the ability to engage meaningfully with the project of attaining understanding of health and illness. The other is the ability to make good predictions about health states, including conditional predictions. These are the Understanding and Prediction Theses respectively. They describe competences with a place in a conception of medicine as an inquiry with a purpose. The Inquiry Thesis *sees* medicine as an inquiry into the nature and causes of health and disease, for the purpose of cure and prevention (that is, for the purpose of removing disease in favor of health). One can make progress in an inquiry without completing it, and this, fundamentally, is what enables medicine to persist despite lack of final success, whether in cure, understanding, or prediction.

We have worked hard on the goal and competence of medicine. Both of these have been framed in terms of health and disease. For a complete philosophical account of the nature of medicine we need to understand health and disease as well. It is time to turn to these.

Chapter 4

Health and Disease

4.1 REMARKS ABOUT THE STATE OF THE ART

The philosophical literature on health and disease uses certain terms differently from their understanding in the broader philosophical literature.[1] It greatly helps to avoid confusion if these differences are understood up front. I will clarify the terms below, but first, it is useful to understand something of the history of each literature—the metaphysical literature, and the philosophical literature on health.

The literature on health saw important developments around the same time as very influential work on the concept of "natural kind," and related uses of the term "natural" such as "natural property" or "law of nature." However, it proceeded either in ignorance of, or in deliberate silence on, these developments (perhaps in part because it did not see much activity at all between the end of the 1970s and the early 2000s). This is unfortunate in that the developments in the metaphysical literature form part of some of the deepest and most useful developments in recent philosophy, resounding through every other wing of analytic philosophy, and representing genuine progress in a discipline where steps forward may be separated by millennia. Who

---

1. Parts of this chapter have been previously published in the *British Journal for the Philosophy of Science* (Broadbent forthcoming).

knows where we would be now if naturalists had spent four decades asking whether health is a natural kind, and not merely whether it is natural.

The developments to which I am referring are sometimes referred to as the "realist turn". They began in the 1970s, especially with the work of giants like Hilary Putnam, David Lewis, and Saul Kripke—thinkers of far greater stature than any philosopher of medicine. In my view it is also genuinely unfortunate that these developments have not made it into the public consciousness, whether because they are so difficult, or because they are dry—their connection to human existence hard to see, or—more likely—because no analytic philosopher who is a capable popular communicator happens to have emerged.

To say that there was a "realist turn" is not to say that everyone became a realist, but rather, that there were developments that overcame what were previously seen as compelling objections to realism, and in turn compelling objections were developed to the previous dominant view, logical positivism or empiricism (the distinction does not matter for us). Likewise it may be asserted that there was a "naturalist turn" in literature on health around the same time, from which the current literature largely springs, and to which it universally makes reference. The phrase "naturalist turn" likewise does not imply that everyone became naturalist overnight, but rather that there were substantial developments in naturalist thinking, and that these appeared compelling to many, both for naturalism and against alternatives.

I now turn to the differences between the meanings of "natural" as used respectively in the philosophy of medicine and philosophy more generally.

In the philosophy of medicine, "natural" can mean one of three things. First, it can mean "objective." Second, it can mean "value-free." Or third, and most commonly, it can mean both. I justify these claims in section 4.2 and merely state them here.

By contrast, the meaning of "natural" in phrases common in metaphysics, such as "natural kind," "natural property," and so forth, mean something like "having real existence in the world regardless of our opinions about them and regardless of what other distinctions we might draw between things." This may seem like a rather complex way of saying "objective," and if so, that is because I have not phrased it adequately. ("Objective" is a weasel word.) The significance of the idea, as applied to kinds, properties, and laws discovered by science, goes beyond mere mind-independence. The idea is that there are real similarities and differences in the world. The similarity between objects of the same mass, for example, is real, because mass is a natural property. Mass features in laws of nature, which relate it to other natural properties such as force. A natural kind is, roughly, a bundle of such properties that come together, and do so in a way that matters. Natural kinds likewise feature in laws of nature: for example, electrons instantiate a natural kind. By contrast, properties such as being made of leather are not natural properties (or at least not very natural—on most views there is a continuum of naturalness). Likewise, kinds such as handbags are not natural kinds. Although they are commonly carried by women and stolen by petty thieves, neither of these is a natural kind either, and there is thus no law of nature linking handbags, women, and thieves.

Note that handbags are still "objective" in the sense that they exist in the external world (to the extent that anything does). It is not their brute existence that depends upon us. It is rather the fact that they are handbags. We could conceivably have manufactured physically identical things for some other purpose, perhaps carrying marbles, and we could have grouped them with other devices for carrying marbles, so that they would just be one variety of marble carrier. The "handbag" way of classifying these objects depends upon us. Whereas this is not so for electrons; they are electrons regardless of what we think of them, and regardless of how we group them with other things—regardless of whether we think they are a variety or proton, for example.

If this makes no sense to you, it could be for one of two reasons. First, it could be because, by instinct, you are not a realist in this sense; and it must be emphasized again that the realist turn did not see all philosophers become realist—rather, it changed the terms of the debate. Second, it could simply be because these matters are complex, and this explanation is too compressed to assist you. I attempt a much fuller introduction elsewhere (Broadbent 2016, 14–30). The interested reader is invited to consult that source, and that is a genuine invitation motivated by my fascination with these matters, not motivated by personal profit—I don't mind if you use a library.

To illustrate the difference between "natural" in the two literatures, consider a phrase like "health is a natural property," something that a naturalist about health might assert. The naturalist about health means that health is objective, or that it is value-free, or both. Whether a given organism or part of an organism is healthy does not at all depend upon whether we think it is healthy, according to the naturalist, and nor does it depend at all upon what value or disvalue we attach to various states of the organism or body part. On the other hand, a philosopher from any other field would understand the phrase to indicate that health is a property really existing in the world. This means more than that it has objective existence. This is most readily illustrated by seeing that the property of being valued or disvalued might have objective existence too. It means something further: that health represents a real respect of similarity between things, that health is related by natural laws to other natural properties, that health "carves the world at the joints," to use the stock phrase.

Naturalists about health do not declare themselves on whether this is what they mean, and common conceptions of naturalism make it unlikely, because on such conceptions health is typically seen as a relation to a population average—typically, the relation of equaling or exceeding. Because population averages can change, and (regardless of potential change) because of the nature of averages (of which there is more than one kind), it is very unlikely that this would meet any criterion for naturalness adhered to by any realist in the metaphysical literature. Thus it seems that the "biostatistical theory of health" famously advanced by Christopher Boorse (see, *inter alia*, Boorse 1977; 1997; 2011), and commonly used as the starting point for naturalist theories of health, would make health rather an *unnatural* property by the lights of the contemporary metaphysical literature.

Thus, in respect of implications of objectivity, the meaning of "natural" is quite a bit thinner in the philosophy of health literature than in the metaphysics literature, to the extent that the most commonly discussed naturalistic definition of health probably renders health a non-natural property by the standards of the metaphysics literature.

However, "natural" also commonly has implications for the status of health in relation to values, in the philosophy of medicine literature (a claim I justify in section 4.2). The implication is that the concept of health is value-free, and thus that the property of health is likewise free from and independent of any value properties. This is a dimension of "natural" entirely absent from the metaphysical literature, which is compatible with the assertion that values are natural properties, or are at least fairly natural.

The sense of "value" at play here is also quite different from that which has been common in the meta-ethical literature for many decades (perhaps centuries). On the common sense of "value," there is no implication either way as to whether values are objective, or subjective, or depend on us (or not) in some other way not captured by either of those two slippery terms. The objectivity or otherwise of values is commonly seen as a central focus of philosophical debate. In the first half of the 20th century, it was widely held that values could not possibly exist independently of us, and various complex theories were devised to explain what values were, and what—if anything—made value judgments true. On many such views, value judgments turn out not to be truth-apt at all, but rather expressions of reactions of like or dislike (emotivism), or universalizable commands (prescriptivism), or something similar. Such positions are called non-cognitivist (as well as anti-realist) because they make value judgment something that is not a function of cognition (understood as relating to beliefs) but rather a function of desires, emotive reactions, expressions of command, and so forth. The expressions of such things are not properly understood as true or false, and thus the fact that nothing in the world answers to such expressions when given a propositional form in natural language does not tell against their meaningfulness or even their validity, in some sense other than truth.

Some such notion of "value" appears to be assumed in the literature on health, where it is commonly assumed that value-dependence is incompatible with objectivity. However, the fact that such complex non-realist, non-cognitivist theories exist should show that this cannot be glibly assumed. Moreover, in the 1970s, moral realism enjoyed a renaissance, continuing to the present day. One important step on this journey was John Mackie's error theory, which rejected both non-cognitivism and realism, coming to the conclusion that moral pronouncements were to be understood literally, but referred to nothing real and thus represented a huge, systematic error on our part. The error theory has few supporters—it seems like a statement of the problem rather than a solution—but critiques of the contortions of the non-cognitivists became more trenchant. Almost every philosophical attempt to take a common form of expression and insist that it really means something different has failed. When one tries to work out the details—for example, how it can be that moral statements can feature in logical deductions just as if they were propositions, if their "logical form" is strictly non-propositional—the complexity becomes insurmountable.

Considerations of this kind led philosophers to look for ways in which evaluative statements could be true, without appealing to some Platonic realm. Sophisticated moral realists look for ways that evaluative assertions, such as "kindness is good," can be literally true, and made true by the natural world. From the perspective of the contemporary meta-ethical literature, it is thus tendentious to see "natural" as having any connection at all with the notion "value-free." From this perspective, health could be a natural property—in the sense of the metaphysics literature—and also an evaluative one, precisely because, if you are a contemporary moral realist, you probably believe that moral values are natural properties, or at least supervene on them in some reasonably natural way.

If this seems confusing, that's because it is. When a "philosophy of" becomes isolated and proceeds in ignorance of, or ignoring, what is going on in the rest of the discipline, the result is at best a divergence of common understanding, and at worst a failure to adopt genuine developments and insights occurring elsewhere, leading to avoidable confusions. In my view, this is the situation in the philosophical literature on health, where two dimensions of disagreement about health have been elided, bundled up, run together, in an opposition between "naturalism" and "normativism." Below I argue that these are not simply opposing positions, nor even extremes on a spectrum, but rather that they disagree along two dimensions, and thus occupy two quadrants of a 2×2 matrix of possible stances on health.

4.2 TWO WAYS TO DISAGREE ABOUT HEALTH

In the philosophical literature on health there are two principal opposing positions: naturalism and normativism. These are typically set
out as unitary and opposing positions in the following fashion:

> Naturalists claim that health and disease are not determined by
> our *subjective* evaluations of a state, but are purely a matter of bio-
> logical fact. Normativists reject this claim to *objectivity* and main-
> tain that health and disease are essentially *value-laden*. [Italics
> added.]

> (Kingma 2010, 242)

On this standard way of setting up the debate, naturalists hold
that health is objective, and value-free, while normativists hold that
health depends upon humans by depending upon their values.

Unfortunately, this standard approach confuses two dimensions
of debate about health: the *objectivity* dimension and the *normativity* dimension. ("Normative" just means "to do with values," where
these may be either moral values, or any other kind of "should" or
"ought"—including the "shoulds" and "oughts" found in our notions
of rationality or aesthetics.)

The objectivity dimension of the debate concerns the extent to
which health facts are judgment-independent. Naturalists believe that
health facts are objective, or at least more objective than normativists
take them to be. Naturalists hold that health facts do not depend on
what we think about them, at least not in any important ways that are
specific to the concept of health and distinguish that concept from
scientific concepts such as mass, organism, or convection rolling.
This is a clear and central part of Christopher Boorse's motivation for
putting health on a firm footing as a "theoretical concept" (Boorse

1977). Clearly, if the recognition of a fact is "a matter of natural science," then the fact in question is, in some central and important sense, objective.

The normativity dimension of the debate concerns the extent to which health facts are value-laden. Naturalists deny that health facts are value-laden. Again, this is clear in Boorse's work:

> On our view disease judgements are value-neutral, which is our second main result. If diseases are deviations from the species' biological design, their recognition is a matter of natural science, not evaluative decision.
>
> (Boorse 1977, 543)

This passage indicates not only the naturalistic commitment to value-free health facts, but also the idea that there is a logical connection between health facts being independent and being value-free.

This perceived connection is apparent in normativist work too:

> By "disease" we aim to pick out a variety of conditions that through being painful, disfiguring or disabling are of interest to us as people. No biological account of disease can be provided because this class of conditions is by its nature anthropocentric and corresponds to no natural class of conditions in the world.
>
> (Cooper 2002, 271)

Cooper holds an opposing position to Boorse's on the objectivity dimension of disagreement: health facts depend on our interests. She is also committed to an opposing position to Boorse's on the normativity dimension: she holds that the *way* health facts depend upon us is through being painful, disfiguring, or disabling, which we single out because they are states we attach some (dis)value to. This normative element is explicit in her definition of disease:

> a condition that it is a bad thing to have, that is such that we consider the afflicted person to have been unlucky, and that can potentially be medically treated.
>
> (Cooper 2002, 271)

There is thus little doubt that there are self-describing naturalists and normativists who align themselves on the two dimensions I describe.

Discussions of the literature likewise tend not to discriminate between these two dimensions of debate. Kingma's discussion of various domains of disagreement between naturalist and normativist begins by casting naturalism and normativism entirely as a disagreement along the normativity dimension. But the way that values are treated in the subsequent discussion, across each of the four domains she distinguishes, makes then non-objective. For example, there is discussion of whether states are "desired" (Kingma 2014, 592); and there is a discussion of "wanted infertility" to illustrate the difference between naturalism and normativism (Kingma 2014, 592). Normative facts of this kind are clearly non-objective, in an important sense—in contrast to the kind of normative fact that a moral realist might assert exists in the world, which *both* escape description in value-free language *and* are objective. Conversely, in the same discussion, the Boorsean naturalist is saddled—no doubt willingly—with the view that a health/disease line cannot "be so blurry that it could be drawn anywhere on the scale, because that would make the health/disease distinction random, and/or too prone to being determined by non-naturalist considerations and/or convention" (Kingma 2014, 595). Thus although Kingma's discussion starts with an apparently clean opposition of naturalism and normativism along the normativity dimension alone, the substance of the discussion clearly assumes that normativism and naturalism also differ along the objectivity dimension. The reference to "convention," in particular, cannot be understood solely with reference to the normativity dimension. Among value theorists, it is by no means common ground that values are conventions; it is only among philosophers of medicine that this assumption appears to dominate.

Similar analyses apply to other discussions that promise to separate the two positions along one dimension alone. An entry in the *Stanford Encyclopedia of Philosophy* divides the debate up between objectivists and "constructivists" (thus selecting the other dimension to the one Kingma emphasizes), but then characterizes constructivism with reference to a reliance not only on judgments but on *value* judgments, which are again implied to be non-objective:

> Although constructivists accept that disease categories refer to known or unknown biological processes they deny that these processes can be identified independently of human values . . . Constructivist conceptions of disease are normative through and through.
>
> (Murphy 2015)

Thus the broad consensus within much of the literature and among many commentators is that seeing health as (reasonably) objective goes hand in hand with seeing health as value-free, while seeing health as value-laden is taken to go hand in hand with seeing health as "subjective" (Kingma 2010, 242), or at least non-objective. Sometimes this connection is stated explicitly and seen as being logical in nature, as in Boorse and Cooper; sometimes it is not stated explicitly but evident from the nature that normative facts are assumed to have, as in Kingma; and sometimes it is stated explicitly but with no logical connection identified, as in the *Stanford Encyclopedia* entry.

Accepting, then, that the objectivity and normativity dimensions of disagreement are not usually separated, the obvious question is whether there are logical connections between the two dimensions, or whether they are logically independent. Elsewhere I have shown in detail that there are no such logical connections (Broadbent forthcoming). For present purposes, I will take this as granted, since it is reasonably obvious. There are many people who believe that facts about right and wrong are objective, and who therefore would deny that value-dependence means non-objectivity. One can obviously believe, then, that health is a value-laden concept (like the normativist) and yet insist that it is perfectly objective, because one believes that values themselves are perfectly objective (contrary to the typical normativist). Such a position is developed by William Stempsey, who even has a book with the subtitle "Value-Dependent Realism" (Stempsey 2000). Conversely, one could insist that disease is not value-laden at all (as does the naturalist) yet insist that it fails to be objective (contrary to the typical naturalist). It is a position of the latter kind that I will end up adopting.

In my view it would aid clarity to move away from the accepted terminology, so that the component commitments are clearer in the positions formerly called "naturalism" and "normativism." Developing Stempsey's terminology, we might rename naturalism *Value-Independent Realism (VIR)* and normativism *Value-Dependent Anti-Realism (VDAR)*. If we adopt this terminology, the availability of a fourth kind of position becomes obvious: *Value-Independent Anti-Realism (VIAR)*. It helps to visualize the situation by setting out a 2×2 matrix as shown in the table below:

	Health is value-free	*Health is value-laden*
Health is objective	Value-Independent Realism (traditional naturalism; e.g., Boorse)	Value-Dependent Realism (e.g., Stempsey)
Health is not objective	Value-Independent Anti-Realism (e.g., the secondary-property view that I will develop)	Value-Dependent Anti-Realism (traditional normativism; e.g., Cooper)

Having set out this framework, I intend to explore the quadrant that is entirely unexplored, to my knowledge, that of VIAD.

There is considerably more to be said about the existing philosophical literature on health and disease. There are other versions besides the versions of naturalism and normativism, along with numerous objections, counters, attempted repairs, further attacks, and so forth. There are also other positions and ideas: hybridism (Caplan 1992; Wakefield 1992; Stegenga 2018, 34–37) which seeks a middle ground between naturalism and normativism (notwithstanding that any recognizable definition of naturalism seems to exclude the logical space for such a position); elimantivism (Ereshefsky 2009) suggesting we just stop talking—or at least philosophizing—about the nature of health and instead talk explicitly about state descriptions and normative claims; comparativism (Schroeder 2013), stemming, for example, from arguments to the effect that the notion of health is fundamentally flawed because it is vague (notwithstanding that the same is true for almost all other natural language predicates); and so forth.

I forgo discussion of that literature here for three reasons. First, it is thoroughly covered in many other cases. Second, any coverage inevitably lands one in further disputes about whether the various positions and objections have been properly characterized, and I do not see the resolution of such disputes as much advancing our philosophical understanding of medicine. Third, that I regard the foundational problem from which the literature springs to be ill formulated, because naturalism and normativism are not the opposing poles of a binary distinction but two positions in a two-dimensional disagreement, as represented in my 2×2 matrix above. This confusion clouds the literature that springs from it; much of that literature ceases to make clear sense when the two dimensions of disagreement are kept in mind. The third reason is the overriding one: it would be a huge effort to review the existing literature with a distinction in mind that was not drawn when that literature was written, and it is not clear why one would undertake such a mind-bending task.

For example, how does one approach hybridism, the view that there are intermediate positions between naturalism and normativism, when one sees two respects in which naturalism and normativism differ? One needs to consider whether the hybridists have in mind differences in objective status only, differences in evaluative status only, or both; or alternatively whether the hybridist is gesturing in some way at the two-dimensionality of the disagreement, so that "between" really indicates the existence of one or both of the other two boxes on the 2x2 matrix; or whether the hybridist is simply confused. One would need to do the same in relation to every objection to hybridism, every response, and so forth. This is a task for the scholastic who seeks to develop every system to its full logical extent, and to compare them. I respect that intellectual task, but it is not the task I have set myself in this book. So I refer the interested reader to recent excellent reviews of the existing philosophical literature (e.g., Smart 2016, 5–43; Sisti and Caplan 2017, 5–15; Stegenga 2018, 25–39), and give myself license to strike out in what I believe is a new direction, and what I hope is a more fruitful one.

4.3 SECONDARY PROPERTIES

VIAR is the view that health facts are neither objective nor value-laden. What possible substantive position could give expression to those two commitments? There may be others, but the one that occurs to me is the view that health is a secondary property.

The primary/secondary property distinction is most famously associated with John Locke's analysis of color (Locke 1706), although similar distinctions had been drawn before. Locke points out that certain properties (he uses the term "quality"), such as shape and mass, are independent of all observation, while others, such as color, depend on observation in an essential way. The wavelength of a ray of light is a primary property, but the color of a reflective surface is a secondary property. Many surfaces reflect electromagnetic radiation, but we have visual apparatus that can detect some of that radiation in certain conditions, and within that we cut the spectrum up in particular ways. To say that a surface is a certain color is to say that, in certain conditions, typical humans would react to the radiation reflected from it in a certain way. The reaction does vary between individuals, but is not something we have control over, at either individual or social level. Some anthropologists have contended that there is some degree of convention in the way the color spectrum is divided up (discussed below). Nonetheless, the phenomenon of seeing color in the first place is not naturally understood as conventional: it really is not something that arises from social conventions, but from the nature of our perceptual systems.

One might object that, on the description just given, colors *are* objective, because the disposition of humans (or even just a single human) to react in a certain way to a certain wavelength of light striking the retina is a perfectly objective thing. This point threatens to lead us into discussions of the nature of consciousness that are largely irrelevant to the philosophy of medicine. Let me briefly indicate two responses to the objection.

First, one might respond that conscious experience is a subjective matter, and the disposition is to have a certain conscious experience. It is a familiar (if subtle) point in the philosophy of mind that there can be objective facts *about* subjective experience, and that this does not show that the subjective is objective. It may be a fact, for example, that I am having a feeling of existential angst at the moment. This may be a perfectly objective fact, in the sense that its truth does not depend on any particular observer's viewpoint. But the experience itself—the thing that the fact is *about*—is nonetheless intrinsically dependent on a certain observer, namely, myself, in that there would be no such thing as my experience of existential angst without my subjectively feeling it. It is at this point that we risk being derailed into a discussion of subjectivity, objectivity, and the nature of conscious experience, which will not advance us much if at all in our understanding of philosophical concerns relating to medicine.

A second reply, posing less of a risk of derailment, is that the disposition of an observer to react in a certain way is all that need be meant by the term "subjective" to draw the distinction in question, provided that the reaction is perceptual or cognitive. This may not seem like a particularly deep or significant distinction; and perhaps it isn't. However, it is enough to support an explanation of color, and enough for the uses that I and others seek to make of it. On this view, a secondary property is simply one that depends on—or perhaps is partly constituted by—the dispositions of observers or thinkers to have certain perceptual or cognitive reactions. Primary properties, by contrast, are objective in the sense that they do not depend on the dispositions of any observer or thinker to react perceptually or cognitively in any particular way.

The extension to thinkers and cognition is probably beyond what Locke or any other 18th-century thinkers had in mind, but this may be to do with their models of mind, which—for the empiricists—permitted very little in the way of what we would now recognize as cognition. On the empiricist view, the mind was cognitively very sparse, capable only of copying what was introduced by the senses, recombining the copies, and inspecting various combinations of these copies for their logical coherence. On this view, there is no clear sense to be made of a disposition to react cognitively; that would merely be a disposition to copy, combine, or examine for coherence, whereas the idea of reacting cognitively is more naturally understood as forming certain beliefs or concepts in reaction to other beliefs we have come to hold or concepts we have come to form. This sort of second-order reaction is difficult to place within an empiricist framework (though perhaps not impossible). I trust that this rather abstract assertion will become clearer below when we discuss causation and, finally, health. In any case, we now credit the mind with a far richer set of capabilities than the empiricists who came up with the distinction between primary and secondary properties. Thus the extension is a reasonable one given the now-common conception of the mind as cognitively rich.

In either of the two senses of the subjective/objective distinction just canvassed, colors are not objective: there is a continuous spectrum of electromagnetic radiation, but the difference between green and red is not important for physics, only for us. Yet neither are colors entirely conventional. In a very famous and widely accepted study, Brent Berlin and Paul Kay argue that cultural variation in basic color terms (like "black," and unlike "charcoal") is limited (Berlin and Kay 1999). All cultures have white and black, and then, if further color terms are added, they follow a predictable order: first red, then yellow or green, and on up to 11 basic colors. Kwame Anthony Appiah discusses colors in exploring the large extent to which humans across cultures have things in common (Appiah 2007, 94–98). The resonance between the secondary-property view of health, and his discussion of colors in the context of limited cross-cultural variation, is one of the things that encouraged me in developing Medical Cosmopolitanism, which we will discuss later in the book.

There is some cross-cultural variation on which basic color terms are in use, and how the spectrum is divided up. In English, "brown" is a basic color term, but in modern Turkish, the most commonly used word is "coffee-colored." However, there are clear nonconventional boundaries on colors. Maybe different societies draw the line between blue and green differently. Maybe different individuals do so within a society. Personally I have the greatest difficulty telling whether others will call certain colors blue or green, even though I can see the difference between them. However, the work of Berlin and Kay shows that basic color terms are added to the vocabulary in a fairly predictable order, and moreover one that is very suggestive: black and white correlate with dark and light, which are obviously important from an evolutionary perspective; red is commonly an indication of danger in the natural world, also suggesting that colors are added because of the evolutionary advantage conferred by distinguishing them; and so forth. But no matter what our cultural background or personal tendencies, we human observers do not see wavelengths that exceed the visible spectrum into either the infrared or the ultraviolet. At the same time, color cannot be the objective property of being a wavelength of electromagnetic radiation. If color were just a wavelength of electromagnetic radiation, then it would exist wherever electromagnetic radiation exists, but there are no colors in the infrared or ultraviolet parts of the spectrum.

Color is something that arises in the course of an interaction between us human observers and the observer-independent world. Bat-observers (at least the blind kind) would not acknowledge colors at all, just as we acknowledge no colors in the infrared and ultraviolet spectra. Snake-observers (the deaf kind) would not have any distinctions between loud and soft sounds, harsh and sweet sounds, and so forth. Presumably, dolphin-observers have sonar-related secondary properties, which we do not acknowledge. What secondary properties there are depends not just on the observed world, but on the nature of the observer in question.

By contrast, primary qualities are not dependent on observers. A red ball with a mass of 5 kg might feel heavier to one person than to another, but it will be 5 kg to both of them, and to anyone, including persons in stronger or weaker gravitational fields. Indeed it need not be 5 kg "to" anybody at all in order to be 5 kg. It simply has a mass of 5 kg. However, the 5-kg red ball is, strictly speaking, red-to-humans. It is not red-to-bats, most of which cannot see, and it may not be red-to-bees, which can see parts of the spectrum that we cannot.

The fact that color is a secondary property enables us to explain some otherwise puzzling things about it. One of the puzzling things about color concerns the fact that colors mix to form other colors. This is familiar, but it becomes strange when one learns that colors "are" particular wavelengths of light, and moreover, that when two colors are "mixed," the result is not a beam of light with a new wavelength, but two superimposed beams with two different wavelengths. Why, then, do we not see two colors at once (indeed, we cannot even conceive of this)? Why do we instead see one color, corresponding to a wavelength of light that—we now learn—does not correspond to either of the wavelengths actually striking the eye, and is not in fact striking the eye at all?

Since we cannot conceive of seeing two colors at once, it helps to compare sound, which is also a wave (of a different kind, in a different medium) and which we also perceive. With sound, we are able to perceive two wavelengths at once. Suppose you play two notes on a piano at the same time, a C and the G above it. You do not hear an E, which is the average of the two wavelengths. You hear two notes, played together—in this case, a pleasantly bracing perfect 5th chord. Now consider what happens if you mix red and green light projected onto a white surface. You do not see a red and green "chord"—that is, you do not see two colors at once. You see one color, yellow, corresponding (roughly) to the color you would see if you projected a light with a wavelength equal to the average of the wavelengths of the red and the green light. Yet in the situation where you project red and green light, the reflected light hitting your eyes is not "averaged out": the light hitting your eyes is of two wavelengths, corresponding to red and green, just as the sound hitting your ears when you play a C and a G is of two wavelengths.

Despite the many differences between light and sound radiation, the explanation for this phenomenon does not depend on those differences, but the way we perceive color. We have receptors in our retina called *cones*, which fire in proportion to the intensity of light of a certain wavelength striking them. These receptors are clustered around three different wavelengths, one each in wavelengths that we perceive as red, green, and blue. A red cone will fire most vigorously in red light, and less vigorously as the wavelength moves away from red. In effect, and very crudely, our visual systems calculate an average of the intensity of the cones' firing, and the color we see arises from that average. This means that we cannot distinguish between a number of physically different situations. We cannot distinguish between (1) a situation where a yellow "wavelength" (to put it crudely—a wavelength does not have a color, as this explanation illustrates) strikes our eye, exciting red and green receptors to about the same extent and blue much less, and (2) a situation where a red and a green wavelength strike our eye simultaneously, which will also excite red and green receptors, and excite blue much less. This is why we see only one color when colors are mixed. It is also why we can reproduce most of the visible spectrum by mixing just three colors (red, green, and blue). It explains color-blindness: those who are red-green color blind have only two kinds of cones, not the usual three. In short, the fact that color is a secondary property explains a number of things that would be extremely difficult to explain otherwise.

The primary/secondary property distinction was originally developed in connection with a distinction between perceptual properties (properties that have a feel, as it were) and non-perceptual properties. However, the essence of the notion of a secondary property is not dependent on sensory experience per se, but on the idea that a property might consist in part of our reactions to the world, as well as the nature of the world to which we react. The idea that health might be a secondary property was prompted in me by Peter Menzies and Huw Price's suggestion that causation might be a secondary property (Menzies and Price 1993).

Causation is clearly not a perceptual property: it is famously something that we cannot see, taste, and so forth. Menzies and Price are motivated by the thought that causation, like traditional perceptual secondary properties such as color and feeling heavy, *seems* to us to be real, and yet is not apparent in the fundamental physics of the universe we inhabit. They suggest that this situation can be explained if causation is something that arises out of our interaction with the world as agents. They endorse an agency theory of causation, and use this to give shape to the suggestion that causation is a secondary property. In their view, causation depends on objective features of the world, but it also depends on us: specifically, it depends on the fact that we are agents with the capacity to move around and intervene in worldly affairs. Michael Dummett put a similar thought by remarking that intelligent trees probably would not have developed the notion of causation (Dummett 1978).

If a view of this sort is correct, then the reason that causation is not discovered empirically alongside mass, charge, spin, and the rest of the fundamental physical properties is that it simply isn't there. But if a view of this kind is right, then causation is certainly nothing like a social convention. It is something that depends on both us and the world; it depends on the place we occupy in the world, the kind of creatures we are, and the nature of the world itself. A world in which, by massive chance, thermodynamic asymmetries do not hold need not violate the fundamental dynamical laws; and in such a world, a causal concept would probably have no grip (Price 1996; Price and Corry 2007). Fires would dwindle into sparks, just as often as sparks would light fires; matches would unlight and jump back into the box; and so forth. We might still see sparks and fires as connected in such a world, but the asymmetry of causation, which is an essential aspect of the concept (Hausman 1998), would be gone. Whatever it would be like to inhabit such a weird world, it seems highly plausible that the inhabitants—whatever they were like—would not share our concept of causation.

The point of this digression is not to establish that causation is a secondary property, but rather to set up a template for arguing that health is a secondary property. Health is unlike color, in that it is not a perceptual property. We do not "see" health or its absence, at least not in the direct way that we see color. In this way, health and causation are alike. They are alike, too, in being the topic of controversy concerning their objectivity. The template for arguing that a property like health or causation is a secondary property is: establish that it is hard to locate an objective basis for some property, which we nevertheless are strongly inclined to treat as objective; then offer an account of the property, according to which it arises from an interaction between us human observers and the world. This accounts for our inability to locate it objectively, and at the same time for its apparent fixity, provided that the interaction between us and the world involves features that are suitably universal, or otherwise such as to be generally unnoticed.

The first part of the template is already fulfilled. It is clear that it is not easy to locate an objective basis for health facts. According to VDR and VIR (aka naturalism), there is such a basis, but even VIR-ists do not typically claim that it is an easy matter to say what it is; otherwise, we should not find it the subject of academic exertion. At the same time, everyone, including VDAR-ists (aka normativists), ought to admit that there is a strong inclination on our part to treat health as an objective property—as something that exists independently of our judgments. Maybe there are VDAR-ists who do not feel this strong inclination, but they must nonetheless accept, as a descriptive claim about fellow humans, that a large number of people talk and act as if health facts were objective. Thus in health we have a property that we naively suppose has an objective basis, but whose objective basis is difficult to describe.

The second part of the template is to offer an account of health that shows *how* it is a secondary quality—on a par with Menzies' and Price's agency account of causation, or with the scientific account of color vision in terms of cones. To this effort I now turn.

4.4 HEALTH AS A SECONDARY PROPERTY

According to Boorse's very well-known account, a body part or process is healthy when it contributes to the survival and reproduction of an organism—its "natural function"—to an extent that is statistically typical or better, relative to other organisms of the same age, sex, and species—a "reference class" (Boorse 1977; Boorse 1997; Boorse 2011).

Very many objections that have been raised against Boorse's account of health (comprehensive listings are given in two rebuttals by that author: Boorse 1997; Boorse 2014). Not all of these are compelling, even where they may appear to have some technical success. Even if the connection between the notions of normality and health are not perfect, there does seem to be a strong conceptual connection between them, as Boorse asserts. It is not a perfect connection; unhealthiness may be the norm, showing that "normal" is a normative and not a statistical notion in this instance. More fundamentally, it seems wrong to seek to disconnect the notion of health altogether from evolutionary biology. Healthy organisms do better than unhealthy ones, as a rule, and this hardly seems a fluke. Health seems to confer an evolutionary advantage. This all tells in favor of Boorse's connection of the notion of health with the notion of species-specific normality, and with the suggestion that this notion has some basis in, or connection with, evolutionary biology. The latter being a science, this lends credibility to Boorse's assertion that health is a "theoretical concept"—that is, a concept featuring (or suitable to feature.

However—and here is the line of objection that I find most persuasive—it is not plausible that health itself is part of the conceptual apparatus of evolutionary biology, nor that it can be derived from that apparatus, as Boorse seems to suggest, by linking the concept of health directly to survival and reproduction of *individual organisms*. Evolutionary biology does not prioritize the individual organism; the unit and level of selection are matters for considerable debate (Okasha 2008), and the organism is by no means the frontrunner in this debate.

Let us explore this concern in more detail. The objections that I find most compelling are those alleging that Boorse lacks "an objective justification for his selection of reference classes" (Kingma 2007, 131). Yielding on this point but insisting that disease is not value-laden puts us into the unexplored quadrant where the view that health is a secondary property lies. Let me therefore elaborate on and endorse the objection that Boorse's choice of reference classes is not objective.

The concern is that there is no proper justification for picking out the reference classes that Boorse selects. Kingma asks us to "imagine there are two candidate concepts for health. One is the BST [Biostatistical Theory, Boorse's view], and one is XST. The XST is exactly like the BST, but has one more reference class: sexual orientation" (Kingma 2007, 131). What empirical reason, Kingma asks, do we have for insisting on BST over XST? To extend the point, what about race? What about ZIP code, make of car (if any), color of hair, favorite color? Is there any reason to pick out age and sex as health-determining, via the reference classes? Is there any *natural* reason?

Boorse's reply is as follows:

> As for choice of reference class, the one that I suggested medicine uses—an age group of a sex of a species—could hardly be a more biologically natural choice. Apart from one detail, the BST's reference class is just one morphological type in the smallest taxon to which an organism belongs.
>
> (Boorse 2014, 695)

Boorse's reply, then, is that his reference classes amount to a natural property, or perhaps to a property with the right degree of naturalness (Lewis 1983); and it is this that justifies their choice. This reply does not answer the right question, however. The question is not "Are the reference classes natural properties?" Naturalness comes in degrees, if it comes at all; and as Boorse points out, a morphological type of the smallest taxon of a species appears to amount to a reasonably natural property. There are objectively significant similarities between property-bearers; these can ground inductive inferences, may feature in high-level laws, and so forth. Perhaps they are not on a par with mass and charge for naturalness, but it would be unreasonable to set the standard for a biologically natural property so high.

Putting it crudely, the question Kingma pushes is not "Are the reference classes objective?" but "Is the *selection* of these reference classes objective?" Thus the question is not whether age group of sex is a natural property (in either the thin "merely objective" sense of the health literature, or the thicker "real respect of similarity" sense of the metaphysics literature). The question is *why this natural property (however understood) should function as the reference class for our health judgments.* It is hard to see how Boorse can do more than just assert that it should, at this point. He could conceivably try to support his position by appealing to the ability of his conception of health to capture actual medical usage, but the fact that there is cultural variation in the content of medicine means that this is a tall order; Boorse's conception permits no deviation in the concept of health. Moreover, even if he can defeat all counterexamples, there is an explanatory question that his theory does not answer.

Compare color. The significance of certain wavelengths of light for a theory of color perception does not arise merely from their being objective. The significance is explained by the fact that we are able to perceive and discriminate between some of them. When we understand this, we see that color is not entirely an objective thing: as a minimum, it depends on our perceptual and/or cognitive dispositions; or more boldly, it depends inextricably on the ineliminably subjective character of experience. Boorse's stance is analogous to the stance of someone who seeks to defend an objectivist thesis about color by repeatedly pointing out that the wavelength of red light is perfectly objective. Of course it is, but the defense fails because it does not explain why we pick this objective property out in the way that we do when we apply the concept "red." Once this explanation is supplied, we understand that color is not wholly objective even though wavelengths are. Similarly, tables and chairs are perfectly objective things, and the difference between a table and chair can be described in physical language; but it has no physical significance. Its significance for us has no physical explanation at all, but can only be understood by appealing to the different reactions we have to, and uses we make of, tables and chairs.

Likewise, until we have an explanation for the connection of health with certain reference classes, health is not wholly objective even if the reference classes are. And even if we do have such an explanation, the explanation may—as in the case of color— show us that health is not, in fact, objective, even if the reference classes are.

The secondary-property framework offers a response to this line of objection. Indeed, there is no biological justification for these crucial claims. That is because health is a secondary property. It is a dispositional property of the natural functioning of organisms, roughly as defined by Boorse, to produce a certain cognitive response in us: the response we express in health judgments. To judge that something (body part or whole organism) is healthy is to judge that its contribution to the organism's survival and reproduction is average or above for the species.

The crucial question for Boorse, the one that I argue Boorse misses in his response to Kingma, is not whether this reference class is natural, or natural enough, but why this natural-enough reference class rather than one of many other possibilities. The secondary-property framework does not answer this question—indeed, it says nothing more than that health is a secondary property, which does not distinguish health from any other secondary property. However, the secondary-property framework enables an answer, to which I now turn.

4.5 EVOLUTION AND THE CONCEPT OF HEALTH

There is nothing objectively special about age group of sex of species. However, it is very special to us, because having average-or-better function for an age group of a sex of a species in relation to the goals of survival and reproduction is the same thing as having better chances in the grand evolutionary game. Assuming, as we must, that our concepts evolved along with the rest of us, it is not hard to imagine how thinking beings might have evolved a concept identifying the contribution that a body part is making to survival and reproduction, or that a whole body is making to the survival and reproduction of the species as a whole. (There is a debate as to whether health is something had by parts or the whole person; I do not have a strong view on this, but suspect the right answer is "both.") With such a concept, we would presumably be better able to devise and then take actions that increase the species' chances of successful survival and reproduction, along with the concept. This is the nature of an evolutionary advantage.

Note that this account does not assume that evolutionary advantage must be cashed out in terms of organisms, and thus that the unit of selection is the organism. A worker ant cannot reproduce. We can still make sense of the notion of a healthy worker ant. The contribution that the sterile worker ant makes to propagation of genetic material is through nourishing a queen that shares genetic material with the worker. The function of the parts of the ant relative to the goal of nourishing the queen are thus what determine the health of the ant, not the direct propagation of genetic material from this worker ant. Such species-specific deviations are readily accommodated on a secondary-property view, because the specialness of the goal by which function is assessed is not a species-independent specialness, but may vary depending on how the species as a whole propagates.

Returning the human realm, relativization to age group of sex of a species is also advantageous, since it allows us to direct our actions appropriately depending on the prospects for an outcome. We do not worry that a baby cannot walk because this is normal for a baby, and in the normal course of things the baby will eventually walk. A concept that did not take this into account would have us wasting time on getting babies to walk too young. Likewise it would have us lamenting the slow sprint of an aged person when there is nothing to be done about it, at least within the Paleolithic context in which our concepts evolved, and thus no point lamenting (although that person may lament it themselves).

Evolution-inspired defenses of a form of VIR-ism (aka naturalism) have been attempted before. For example, Karen Neander and Benjamin Smart both seek to clarify the notion of natural function by reference to the function for which a body part or process evolved (Neander 2006; Smart 2016). Similarly, Mahesh Ananth's book-length defense of an "evolutionary concept" of health defends the idea that natural function can be explicated with reference to the goals of survival and reproduction (Ananth 2008). However, these approaches differ substantially from the sketch being advocated here. They attempt to explain natural function in terms of the contribution that a body part or process made, historically, to the survival and reproduction of organisms of the species. This is quite a different use of evolutionary explanation, since it aims to clarify the sense in which functions may be natural. My use, on the other hand, makes no claims on naturalness, but rather seeks to explain the fact we have a concept picking one equally natural set of functions out rather than another, in terms of the contribution that being able to pick them out would make to survival and reproduction.

I have no quarrel with an evolutionary account of natural function. The possible account that I have sketched is compatible with such accounts. But I do not think that, on their own, accounts such as those of Neander, Ananth, and Smart are an adequate response to the explanatory question posed by Kingma. They may show how the functions tracked by the health concept can be natural, but they do not explain why our health concept should track *these* natural functions. Showing that a function is natural is one thing; showing we have a reason to care about it is another. An evolutionary explanation of the health *concept* is one that shows how it helps with survival and reproduction; and a concept tracking these things would, one imagines, be relatively easy to explain, since it is relatively easy to imagine that it might confer an advantage in survival and reproduction.

I must stress that this evolutionary account of the concept of health is just a sketch. I have not made the detailed empirical argument that would be needed to fill the sketch out. A proper development would require a thorough evaluation of the kind of evolutionary advantage that a health concept might endow, and would need to guard against the risks of ad hoc explanation and the adaptationist fallacy that beset every evolutionary explanation of an actual trait, especially of social and cognitive ones.

Nonetheless, the sketch illustrates how a naturalistic account of health might be served by the secondary-property approach. It is compatible with an evolutionary account of our coming to have the concept of health, and that is thoroughly naturalistic. Perhaps the deep problem with Boorse's view is that the naturalism does not go deep enough: at some point, he just insists that certain goals relativized to certain reference classes are natural, without explaining how. If we accept that it is we, and not the world, that pick out the reference classes, then we can offer an explanation of why we do so. Contrary to the normativist, this explanation need have nothing to do with values: it may be entirely to do with the fact that creatures with this kind of concept survive and reproduce.

There are many more objections to naturalism (Gabbay, Thagard, and Woods 2011; Smart 2016) and I reiterate that I do not pretend to address them all. I do, however, believe that by distinguishing commitments to objectivity and to being value-free, it is possible to show how objections to the objectivity element of naturalism can be conceded without thereby necessarily conceding that health is value-laden. A secondary-property view shows how this is possible, and an evolutionary account of the health concept would—if developed successfully—enable a value-free explanation of why that concept should track certain goals relative to certain reference classes. This would be thoroughly in keeping with the spirit of naturalism: one can concede that health facts are not objective without conceding that they are value-laden, and the kind of account that one can offer for the dependence of health facts on us may be thoroughly naturalistic.

4.6 HEALTH IN DIFFERENT TIMES AND PLACES

I have been at pains to emphasize that the content of medicine differs hugely from time to time and place to place. The concept of health is either part of the content of medicine, or else heavily influenced by medical opinion. It is quite common to ask a doctor whether something you are experiencing is healthy. However, the analysis of health that I have presented appears to give it some properties that are definitely not universally recognized. It is naturalistic, and subject to an evolutionary explanation. Not all medicine is naturalistic, and certainly not all medicine is informed by the theory of evolution, since that theory has not existed for most of the history of medicine. How does the analysis of health that I have offered apply across different cultures?

It is important to distinguish between what, in a different era and a different philosophical specialty, would have been called *intension* and *extension*, or—harking back still further—*connotation* and *denotation*. "Creature with a heart" and "creature with kidneys" have quite different connotations, or intensions, or—roughly—meanings. Yet the phrases refer to the same actual things (because all creatures with a heart have kidneys, and all creatures with kidneys have hearts). My analysis of health is not an analysis of the concept of health, giving its connotation, or meaning. Rather, it is meant to be an analysis of the extension, or denotation of "health": that is, of the thing itself, to which the concept refers. I am not trying to say what "health" means; I am trying to say what health is.

In this respect my approach differs significantly from much of the existing literature, which often adopts the approach of conceptual analysis. As I have already indicated, and discuss at greater length elsewhere (Broadbent 2016, 136–54), I do not think this is a fruitful approach. The secondary-quality framework is not supposed to indicate that the meaning of the word "health" includes anything about secondary qualities—no more so than a secondary-quality analysis of "color" is supposed to indicate that "color" means "secondary property." Likewise, the evolutionary component of the account of health is not supposed to indicate that "health" means "that which promotes chances of survival and reproduction." On the contrary, we have many concepts whose place in our conceptual apparatus is probably explained by improving our chances of survival and reproduction, but which do not *mean* anything like that. Consider the concept of disgust, a commonly cited example of a concept that is universal in human societies, and yet which is populated by different objects at different times (the extent of the difference being a matter of study). Consider, even, the concept "sexy": clearly not equivalent to "has high chances of surviving and reproducing," and yet surely part of a conceptual family whose main and general function is to articulate the sex drive, in ways that make that drive generally more likely to be successful.

It is thus no objection to my account that the meaning of health varies as widely as does any other part of the content of medicine.

Whether it is explained in terms of the flow of *qi*, the balance of humors, or biological function, the concept is recognizably that of health. The reason it is recognizable as such, despite having such wildly different meanings in different places and times, is because it picks out *broadly* the same thing. Not exactly; just as there is cross-cultural variation in color terms, there is cross-cultural variation in what is counted as healthy. But, as Appiah emphasizes (Appiah 2007, 87–98), the reason we even call this cross-cultural variation in the same concept is that there is so much in common in the first place. That commonality, I suggest, is broadly the picking out of states that promote the survival and reproduction of the species (not necessarily the individual) and thus confer an evolutionary advantage. It is, after all, very likely that such a concept would come to dominate, since it is so obvious that it would be useful to a species that possessed it.

4.7 DISEASE

Disease is more than just the absence of health. Every medical tradition has more to say about disease than that it is health's absence. This "more" can take different forms. In contemporary Mainstream Medicine, there is a distinction between two meanings of "disease," one being simple absence of health (as in "the liver is diseased"), and the other indicating a particular *kind* of absence of health (as in "cholera is a different disease from dysentery"). In Mainstream Medicine, it is a matter of debate what makes the difference between different kinds of disease. During the late 19th and early 20th century, germ theorists held that every disease was caused by the invasion of a microorganism, and thus that diseases could, ultimately, be carved up into kinds according to which species of microorganism was responsible. However, it became clear that this was true only for some diseases. Others are not caused by microorganisms at all; and even those caused by microorganisms might not be so simple as we thought, when we consider interactions, asymptomatic infections, and immunity.

In other traditions, the notion of disease does not lend itself to a kind structure at all. For example, in Hippocratic medicine, diseases were seen as basically unique to the individual sufferer, since they arose from an imbalance of humors. Different individuals could suffer similar imbalances, meaning that medical generalizations relating to care and treatment were still possible. But there was no separate "disease entity"; the disease was fundamentally a property of the person, specifically an imbalance of humors. Within such a framework there is much less inclination to say that two people suffer from the same disease, or the same kind of disease, but rather, that their diseases are very similar or the same. The disease is a property of the person, in Hippocratic medicine, whereas in Mainstream Medicine it is often (though perhaps not always) conceived of as a separate entity from the person.

I believe that there is much to be said about the concept of disease, and particularly about how diseases can and should be classified into kinds. However, this does not fall within the scope of understanding the foundations of medicine itself, but rather in the normative project of trying to direct and improve medical concepts so that they are more effective. For our purposes, those of saying what medicine is, the secondary-property account of health and its evolutionary explanation show how and why the meaning of the health concept (its connotation, or intension) can vary greatly from place to place and time to time, while still referring to the same thing. Similarly, the notion of disease varies, and for the same reasons. In trying to understand why health sometimes fails, people have come up with different alternative explanations, and other than the constraint of seeking to understand states that are not conducive to survival and reproduction (of the species), these efforts are limited only by imagination.

Of course, their success as explanations can be evaluated by the success of the corresponding systems of medicine at achieving the goal of cure. I do not mean to suggest that all notions of disease are equally correct: as we shall see, I am not a relativist. But we are now encroaching on questions that are for the second part of the book.

4.8 CONCLUSION

Health is the collection of human states that are generally conducive to the survival and reproduction of the species. It is a secondary property, because the grouping of these natural properties and their salience arises from our reactions to it, and not from any salience conferred, somehow, by nature itself. Like color, what counts as healthy can vary from culture to culture; but like color, there is a strong limit on this variation. That limit arises because there are facts about what improves chances of survival and reproduction, and these facts favor the concepts that promote them. It is obvious that a concept of health, if it picks out health, will generally favor survival and reproduction; indeed, medicine, whose goal is to remove disease in favor of health, probably plays a role in the recent increases in human population, just as disease probably plays a role in keeping it in check.

In this chapter, we have touched on the question of whether health and disease are relative to a point of view. In doing so, we have veered even more closely than before into questions about the attitude we should have toward medicine, and its concepts. Rather than jump straight in with that question, I want to take a systematic approach, considering various ideas about what attitude we should have toward medicine, starting with those that are most well worked out, and most influential. This is the task of the second part of the book.

WHAT SHOULD WE THINK OF MEDICINE?

Chapter 5

Evidence-Based Medicine

5.1 EBM AS AN ATTITUDE

What should we think of medicine? More fully, what attitude should we adopt toward it? The question faces us whenever we have a decision to make about whether to seek medical treatment in the first place, then when deciding what kind of treatment to seek, and finally when deciding what to do with the practitioner's prescription or recommendation. I have been suffering from a sore neck recently, and eventually decided to go and see a physiotherapist. I could also have gone to see a biokineticist, chiropractor, osteopath, general practitioner, orthopedic surgeon, acupuncturist, Reiki healer, *sangoma*, or someone else again. I could also not have sought medical attention in the first place. The physio did some massage, suggested a follow-up, and gave me some exercises. I could do the exercises, or neglect them; I could go to the follow-up, or cancel. My decisions about all these things are guided by an attitude toward the medical treatment I receive and expect. I assess my treatment. In this chapter, the attitude toward medicine that I am interested in is the attitude that enables me to do so.

The question is a normative one, but not one of normative ethics (though there are ethical dimensions, as we shall see later in this chapter). It is a question of normative epistemology, and of practical action. What should we believe about medicine, and what should we do in relation to it?

This question cannot be answered unconditionally. It depends on medicine satisfying certain conditions. For example, if your view is that some medical tradition—let us suppose Mainstream Medicine—is fundamentally sound, despite being wrong or ignorant on points, and therefore should be trusted, and its prescriptions followed, then you presumably mean Mainstream Medicine *done properly*. You do not mean anything that comes out of a doctor's mouth, regardless of whether she is drunk, suffering from a bout of schizophrenia, or in the pocket of a pharmaceutical company. Accordingly, an attitude to medicine is sometimes expressed in part by painting a picture of how medicine ought to be, with the attitude left almost implicit. Such is the case with Evidence-Based Medicine (EBM), which includes both a prescription for medicine and a prescription for the patient—an attitude that the patient ought to adopt. EBM expresses strong views about how medicine has been, is, and ought to be; and it suggests that patients adopt an attitude that is both critical and trusting: critical, in ensuring that the standards it prescribes are adhered to, and trusting when they are. It is therefore a good place to begin a discussion of the attitude we should take to medicine, which, as I have indicated, cannot be fully separated from the question of what medicine *ought* to be like.

5.2 A PRESCRIPTION FOR MEDICINE

EBM is the most provocative and influential development in, and in some ways against, Mainstream Medicine of recent times. Depending on your point of view, you may see it as a clarion call to rational, pragmatic, and ethical medical practice, at long last challenging a system that protects the egos of those occupying high positions in the medical hierarchy; or as a dangerous, muddled, or largely vacuous hullabaloo whose primary purpose was in fact to serve the egos of a few ambitious individuals (sometimes known off the record as "The Silverbacks"). Likewise, depending who you ask, you may hear that EBM has exerted either a huge or a negligible effect on how medicine is actually practiced, and that this effect has either been largely positive or largely negative.

Part of the reason for this is that EBM is fundamentally a *social phenomenon*. By this I mean that the methodological points at issue could have been considered by the medical profession, and adopted, rejected, or modified, without anything called "EBM" ever coming into existence. At least, they could from a *methodological* point of view; whether the Mainstream Medical profession would have been able to do so is another question—and a sociological one.

EBM is a movement within the medical profession, beginning in the mid-1990s (Sackett and Rosenberg 1995). EBM starts with an idea that we have already encountered in Chapters 1 and 2, that ineffective or harmful medicine has dominated the history of medicine. However, EBM extends this view to include contemporary medicine: EBM's impetus comes from the conviction that even contemporary medicine was often wrong in its prescriptions.

Thus in the foreword to a book that might be regarded as some sort of culmination of the EBM movement, *Testing Treatments*, we find this:

> the triumphs of modern medicine can easily lead us to overlook many of its ongoing problems. Even today, too much medical decision-making is based on poor evidence. There are still too many medical treatments that harm patients, some that are of little or no proven benefit, and others that are worthwhile but are not used enough.
>
> (Evans et al. 2011, xx, in a foreword by Ben Goldacre)

As we will see in Chapter 5, there are striking similarities between this sentiment and the views of medical nihilists. The difference is that EBM believes that there is an answer to this problem: the explicit and systematic use of good published research in order to make clinical decisions. Obviously, a lot of rubbish is published; thus the crux of EBM is its idea about what makes research good. And here, the resounding answer is: randomized controlled trials (RCTs) are the best kind of empirical study.

There were some high-profile episodes that supported this view. Hormone replacement therapy (HRT) was thought to be protective against coronary heart disease (CHD) based on a review of all the evidence available from observational studies, even when all known or suspected confounders were adjusted for. However, a famous RCT apparently showed that this was not the case. Whether these high-profile cases warrant the reaction of EBM is an interesting question; but in any case, the conclusion drawn was that medicine as a whole needed a revamp. It relied too much on wisdom handed down from teacher to pupil, and needed an injection of "skepticemia."

EBM is a movement within medicine, but it is also an attitude toward medicine. The two go together, because the movement says what medicine ideally should be like, and the attitude we should have toward medicine is determined by the extent to which it attains the ideal. According to EBM, we should believe the pronouncements of medicine to a greater or lesser degree, depending upon the evidence base for those pronouncements. This might sound banal, but the meat of EBM lies in its views about what counts as evidence.

The central ideas behind EBM might be summarized as follows. Medicine has often failed to base its prescriptions upon good evidence of their effectiveness. This ought to be corrected, and clinical decisions ought to be based upon the best available evidence of effectiveness, with confidence in those decisions tempered by an assessment of how good the best available evidence is. The goodness of evidence depends on two factors: first, how directly it demonstrates the effectiveness of the treatment at hand; and second, how resistant it is to bias and confounding. EBM arranges types of evidence into a pyramid according to these two factors. The pyramid is a lexical ordering, with higher kinds of evidence trumping lower kinds, meaning that very small quantities of higher-level evidence can outweigh very large quantities of lower kinds. Many of these statements have been qualified by various authors at various times, but they are sometimes defended as good rules of thumb for practical purposes, and have also been relied upon for rhetorical ones. In particular, the trumping character of RCTs is relied upon in the HRT narrative above. Also, the figure of a pyramid rather than a ladder or pillar suggests that there may and often will be much more of the lower-quality evidence available, and that this does not matter to the primacy of the more elevated kinds of evidence.

The exact contents of the pyramid vary a bit from statement to statement, but there is much in common—in particular, the underlying notion of a lexical ordering (pyramid) of evidence. At the very top are meta-analyses, which consist in analyses of the evidence already provided. However, these are not empirical studies; the highest empirical studies are RCTs, followed by cohort studies, followed by case-control studies, and at the bottom, expert clinical judgment and reasoning from laboratory scientific knowledge ("bench science").

Above I identified two factors determining the goodness of evidence in the EBM framework: directness of bearing on the treatment in question, and resistance to biases and confounding. The two factors identified above bearing on the goodness of evidence are clearly operative in the pyramid. Expert knowledge may bear very directly on the treatment at hand, but it is of little value because it is admittedly subjective and unsystematic, and hence vulnerable to bias. Laboratory science is the other way round. It is *better* controlled than an RCT, and less vulnerable to bias and confounding. But the evidence it provides for effectiveness is indirect: EBM places very little faith in reasoning from the way something ought to work, to how it will work.

At the next level up, we have case-control evidence, which may bear directly on the treatment, and is at least systematic empirical research of some kind (unlike unsystematic clinical experience). However, it is open to some well-documented biases. Case-control studies are *observational* studies, meaning that no experimental intervention is made; the investigators merely observe what happens and seek to draw conclusions on that basis. In a case-control study, cases of an outcome of interest are identified, and then the prevalence of the exposure of interest is compared to the prevalence among a control group. Thus one might find out what proportion of lung cancer cases in British hospitals were heavy smokers, and compare this to a control. Identifying a suitable control group is not easy. Hospital inmates suffering other diseases are one option; persons matched one to one for age, sex, and other obvious confounders provide another option; and a random sample from the general population provides a third. Sometimes, all these options might be used, to improve the robustness of the study (Peacock and Peackock 2011, 24).

Case-control studies suffer from various difficulties. There are some kinds of bias to which they are peculiarly vulnerable. In addition, they are logically curious: they tell us how many lung cancer sufferers smoke, but not directly how many smokers get lung cancer. Intuitively, the former is a guide to the latter; if we discover that lung cancer sufferers more often smoke than non-sufferers, we will naturally infer that smokers more commonly get lung cancer than non-smokers. But logically and mathematically, further assumptions are required to get us from one to the other (Broadbent 2013, Chapter 2). Most fundamentally, they cannot easily separate out the effect of the many differences between cases and controls from the effect of the exposure of interest. The reason that somebody ends up as a case might also explain the putative exposure; a genetic trait may predispose toward smoking, and also toward lung cancer, and in that case the case-control study of smoking and lung cancer would show a correlation for that reason and not because smoking causes lung cancer. This argument was used to resist the conclusion that smoking caused lung cancer, when case-control studies were the only kind of evidence available for that conclusion.

Cohort studies do better against some of these biases, so are a level higher on the EBM pyramid. In a cohort study, a "cohort" or study population is identified and followed over a period of time, potentially a very long period. The cohort is assessed for exposure(s) of interest and outcome(s) of interest. Cohort studies offer more direct evidence for causality because they permit the direct calculation

RCTs are the best (of the empirical studies) because randomization protects against bias and confounding: or, more exactly, it enables an exact quantitative estimate of the probability of error. In an RCT, a study population is split into two (or more) equal parts at random. The treatment is allocated to one group and a placebo to the other. Ideally, the trial is double-blinded, so that neither the investigators nor the trial participants (nor anyone else directly involved in the trial) know which group is receiving the treatment and which is receiving the placebo.

Despite their name, RCTs are not truly controlled, in the sense in which controlled experiments are controlled. The term "control" means, obviously, that the situation is under the control of the investigator. In a controlled experiment, this allows the investigator to adjust one variable while keeping all others the same in the "control" apparatus, and then comparing the two. Subsequent differences are attributed to the antecedent adjustment, the inference being that there were no other potentially causal differences between the two set-ups. Of course, there might have been undetected ones, so replication is important; and there are always many irrelevant differences: test tube

1 might be to the south of test tube 2, for example. Sometimes these supposedly irrelevant differences are in fact relevant. But in the ideal controlled experiment, and in many actual controlled experiments, we have good reason to think that all relevant factors—all that could potentially influence the outcome—are controlled.

RCTs are not like this. The differences between people typically do have the potential to affect the outcome of the treatment. And, except for some crude stratifying by sex, age, and so forth, they are not controlled at all by the experimenter. The big idea behind an RCT is that randomization can act as a surrogate for true controlling; it can achieve an approximation of what can be achieved by actual controlling. Some trial participants live to the south of the river, some to the north; some have better education; some are vegetarian; some have undetected tumors, and so forth. If these potentially causal factors are distributed at random, then, on average, their effects will cancel out. Given enough people, it is highly unlikely that a random allocation will put all 1,000 participants who live south of the river in the treatment arm and all 1,000 who live north of the river in the control arm. It is possible, of course; hence my point that this is not true controlling. But it is very unlikely. And more to the point, we can calculate *how* unlikely this is, and thus achieve a quantitative estimate of the probability that our finding arose from some undetected causal factor being unevenly distributed between treatment and control group. The value standardly used to express this probability is known as the p-value, which is a probability (a number between 0 and 1) telling us how probable it is that the outcome would have come about by chance (to put it roughly).

EBM proposes that clinical decision-making be based on the best available evidence. RCTs are meant to "trump" observational studies, and the latter are significantly worse, in EBM's eyes, than RCTs. Much of the criticism of EBM has concerned RCTs, pointing out ways they can go wrong, and things they cannot do. We will cover some of this criticism in the remainder of this chapter. It is easy to wonder, sympathetically, whether the criticism is fair: does EBM really place so much faith in the RCT, and so little faith in anything else? The answers are "yes" and "yes." The EBM movement sees the "ascendancy" of the RCT as a "fundamental shift" in clinical evidence (Sackett and Rosenberg 1995, 620). Clinicians are advised to "stop reading . . . and go on to the next article" upon discovering that a study is not randomized (Sackett 1997, 32).

Reasoning from biological, chemical, or anatomical knowledge is even more suspect, in the EBM picture. Indeed, EBM might reasonably be thought of as an empiricist turn in medicine, given its strong suspicion of conclusions derived by reasoning from other knowledge. Conclusions that have not been subjected to direct empirical test are to be treated with severe skepticism, according to EBM.

The famous pyramid diagram that is often used to encapsulate EBM ideas (RCTs at the top, expert opinion and bench science at the bottom) reveals something important about EBM: it is a social movement as much as a methodological one. The pyramid has no basis in methodological research and no parallel in any other science. So far as I know, no scientist has ever proposed classifying evidence by types in this way, and arranging it in a hierarchy, prior to EBM. My inference, then, is that the pyramid draws its inspiration from a social structure, namely the professional hierarchy of medicine, and further that its inclusion is meant to serve a social function, that of challenging the medical hierarchy. This would explain much that is otherwise confusing, such as the strong headline claims of superiority of certain kinds of evidence, which are then considerably nuanced in discussion or practice. There is also the apparent neglect of the methodological principle familiar to science and detective fiction alike that one must consider all the evidence, and neglect nothing, no matter how trivial it may seem. This was a favorite point of Sherlock Holmes', the creation of a medical man (Arthur Conan Doyle was a doctor), and has been explored in rather less appealing ways by many other methodological thinkers (perhaps most notably Carnap 1947).

Even if my speculation about the function of the evidence hierarchy in EBM is incorrect, it is undoubtedly true that EBM included significant social and professional goals. It is worth bearing this in mind when considering the various criticisms of EBM, and the level of frustration that sometimes builds up on both sides. Critics and proponents have sometimes talked past each other, and one reason for this may be that the movement and its pronouncements sometimes had social or professional goals rather than strictly scientific or methodological ones. It was an attempt to change medicine, and changing any social institution requires more than simply stating a truth and the reasons for it.

5.3 A PRESCRIPTION FOR PATIENTS

In theory, EBM takes authority from experts, and puts some of that power into the hands of patients. Of course, practice is a different matter, but in theory this is one effect of the evidence hierarchy. Patients can, in principle, inform themselves of the evidence base for a proposed course of treatment. In an ideal EBM consultation, the doctor will consult the evidence in relation to both achieving a diagnosis and recommending a treatment, and the patient will be involved in this process. The role of the physician becomes that of a guide to the evidence and a translator for it. The patient can further inform herself using the internet, and one of the most dramatic innovations of EBM is the Cochrane Collaboration, supporting a huge, publicly accessible database of evidence relating to effectiveness of treatments. If there is no high-quality evidence, then the doctor's own opinion based on clinical experience or anatomical and biological knowledge might be the best that the doctor–patient team has to go on. But this is the last resort, not the first approach.

EBM has sometimes been criticized for excluding the patient, as we shall see below. However, it is important to see that the ideology of EBM is in theory supposed to promote "patient power," and confine the epistemic authority of the doctor. "Skepticemia" can and should infect patients as much as doctors; when both are infected, the pair becomes a problem-solving team, ideally, leaving egos, social power relations, and anything else likely to cloud a rational judgment at the door. The doctor guides and translates, but the patient decides where to go (this is where values play a role), and can feed meaningful information into the process by doing her own research, correcting reasoning errors, and so forth. It is all very "second Enlightenment," and Enlightenment values are sometimes espoused by proponents of EBM. Such harking back to a bygone age of reason is based on a rosy image of the past, rather than any detailed historical research (akin to the tendency of some Americans to hark back to the Founding Fathers). But it is the image rather than its accuracy that is important for us. That is an image of dispassionate yet kindly rationality, fairness, confidence in the reasoning and investigatory powers of the human mind mixed (not entirely comfortably) with open skepticism when those powers approach their limits. Better to admit ignorance than profess knowledge without a base, but better to seek to assess for oneself others' claims to knowledge than to accept them on authority. Like-minded people won't object, and those who object should be treated with suspicion.

The promotion of patient power is thus within strict limits. It is not a promotion of patients' wishes or views. It is not that patients are given more authority, but rather that doctors are brought down. Both are subservient to the evidence.

There is no denying the appeal of EBM. It is surely better to rely on better evidence than worse, and although the hierarchy is novel, it is justified by reasonable methodological considerations. If randomization is possible, it is better to do it. If one wants to know whether a given expert is really an expert, and it is possible to gather some empirical evidence relating to the claims that the expert makes, then this is a preferable means of assessing expertise than simply asking another expert. And so forth for the other relationships in the hierarchy. Nonetheless, EBM has faced trenchant criticism, of varying degrees of cogency. I now turn to a concise survey of some of these criticisms.

5.4 TRIALS AND BIASES

As indicated above, there are two aspects to the goodness of evidence in EBM: resistance to bias, and specificity to the contemplated treatment. Critics have questioned the ability of EBM's prescription for medicine to meet either challenge. In this section, I focus on the bias, and in the next, on specificity.

RCTs can be performed poorly in practice, or can go wrong by accident. There is plenty of evidence that blinding fails quite often on RCTs, although quantifying this is hard; and when blinding fails, randomization may also be threatened. There are anecdotes about investigators deliberately breaking the blinding—holding envelopes up to the light, rifling through filing cabinets, and generally going to unseemly lengths to discover which groups are receiving which treatment. Such actions may be sinister and corrupt, but they may also arise from simple curiosity: sometimes investigators may just want to know which pill is the placebo (Howick 2017, 123–25). If a drug is effective, or has dramatic side effects, then knowledge (or suspicion) may also arise quite accidentally in the mind of investigator, physician, or patient, from the effectiveness of the treatment. In the first trials of Viagra, originally intended as a drug for hypertension and angina pectoris (Kling 2998), there was no doubt that the drug was having some effect, even if not the one that was originally intended. Doctors are also meant to keep an eye wide open for bad or dangerous effects. This scrutiny might well result in patients, physicians, and investigators being able to hazard a very good guess, by the end of the trial, as to whether the substance they had been taking was effective.

Even if RCTs are performed with scrupulous perfection, there is no escaping chance. Statistically speaking, 1 in every 20 RCTs showing a correlation with a p-value of 0.05 arrives at that result by chance. Although large trials make the operation of chance seem intuitively implausible, intuition is a poor guide to statistical matters. Moreover, a p-value of 0.05 is the least demanding standard applied in any science; physicists searching for signals in a particle accelerator (where the numbers are many times larger than those in even a large trial) standardly use $p < 0.001$ (or 1 in 1,000 chance of a false positive). A famous empirical demonstration showed an apparent effect of retroactive, interdenominational intercessory prayer in a large randomized trial (Leibovici 2001). The retroactivity is important; even God would surely struggle to know that, two years later, in some academic's office, someone would mutter a short and vague incantation in favor of a couple of thousand randomly selected unknown persons whose diseases had occurred some years earlier. Error is part of science, and any reasonable proponent of EBM will admit that even RCTs are subject to the vagaries of chance.

Chance can be ironed out to a large extent through replication, meta-analysis, and systematic review, even if it never disappears completely. These tools are less potent against three important systematic biases: publication bias, financial bias, and pill bias. Publication bias is sometimes known as the desk drawer problem: if you do a study and find nothing—no correlation—you are less likely to publish it than if you do find a correlation. (You just put it in the desk drawer.) By its nature, publication bias is very hard to quantify, but it is obviously important because it skews the odds that I have just described. If 20 trials are done, and there is in fact no correlation to be found, then, probably, one of them will detect a correlation with p-value 0.05. If publication bias operates, this one has a much better chance of being published than the others. Suppose there are then attempts to replicate, and the bias operates again here, so that failures to replicate are less likely to be published than successful replications. The result might be a mix of results, some showing a correlation and some not, with the mix looking much more favorable to the existence of a correlation than it would if all the unpublished work that does not find a correlation were included. The "system" does seek to remedy this; failures to replicate *are* published, as are findings of no correlation. Nonetheless, because the problem is hard to quantify, it remains troubling. The extent to which the theoretical probabilities of chance error are skewed by unrepresentative publication patterns is an unknown unknown.

Financial bias, on the other hand, is well-documented, although its extent is hard to assess because it may be either unconscious or clandestine. Financial bias is the tendency of studies to find results that favor the interests of those financing the study. Usually, those interests are themselves financial, though they could be religious, political, or just plain cranky. Financial bias operates to affect the initial choice of an object of study. The existence of what are commonly known as "neglected tropical diseases" provides an illustration. These are diseases whose sufferers are mainly poor, in poor or poorly managed nations. For obvious commercial reasons, a pharmaceutical company will need to establish that a market exists and that a commercially viable product might be produced for that market. If the market does not exist because people who suffer a certain disease are typically unable to pay for a treatment, or if a drug cannot be patented, it will make no commercial sense to invest the large amount of money that drug development requires (Broadbent 2011a).

Financial bias also operates on the probability of finding a positive result. It is very hard to detect in particular cases; that is, it is very hard to demonstrate that financial bias has affected this or that trial. However, at the population level, the effect is clear. A well-cited, if now rather old, systematic review estimated that industry-funded trials are four times as likely to produce positive results (Lexchin, Bero, and Djulbegovic 2003). The impact of this problem on the biomedical research system is potentially huge, because much biomedical research is funded by industry. The journalist and commentator Ben Goldacre puts the proportion of trials funded by industry at 90% (Goldacre 2011, x). What is more, financial bias particularly affects RCTs, because RCTs are very expensive.

A third important kind of bias concerns the kinds of treatments that can be investigated. There is no generally accepted name for this, although it has been discussed in the literature. I call it "pill bias," because treatments that come in the form of a pill are ideal for RCTs. It is usually easy to make a placebo pill that resembles the treatment pill closely enough to be indistinguishable on cursory inspection. However, some interventions cannot be tested by RCT for ethical reasons, others cannot be tested for practical reasons, and others cannot be tested by RCT even in principle.

Consider this remark by Ben Goldacre in the foreword to the book *Testing Treatments*:

> any claim made about an intervention having an effect can be subjected to a transparent fair test.
>
> (Goldacre 2011, xi)

Assuming that a test is some form of study that involves an intervention on the part of the investigator, this claim is an overstatement; or, to put it bluntly (as Goldacre would), it is false. Not all treatments can be subject to every kind of test; and of those that can, not all are susceptible to a "transparent fair test," depending on what the terms "transparent" and "fair" mean.

Interventions causing harm cannot be tested by RCT with a clear conscience. To take the famous example, this means that a standard RCT to test for a correlation between smoking and lung cancer cannot be run. Practical difficulties with RCTs can arise, for example, when a treatment has very onerous compliance requirements. Thus, even if in theory one could conduct a randomized trial on the effect of certain dietary habits, with carefully constructed placebo foods with indistinguishable taste and texture, such foods are not available in supermarkets and would be exorbitantly expensive to supply. High-quality nutritional RCTs may be notionally possible but are often simply not practical.

Some treatments are even impossible to test in principle without relaxing some of the requirements of an RCT. Some treatments simply cannot be concealed from the patient, the investigator, or both. Exercise, for instance, cannot be tested without participants knowing at least what exercise they are doing. Physiotherapy, massage in general, acupuncture, osteopathy, and chiropractic are likewise impossible to conceal from the participant. The best one can manage is a comparison between different apparently similar treatments: thus one could test acupuncture against the random insertion of needles, or physiotherapeutic massage against other kinds or random massage. Surgery, too, is often impossible to conceal; in theory one could cut a hole in a patient, do nothing, and then tell them the operation was performed, but this is clearly wrong, and moreover sometimes not even possible. Consider the extraction of a tooth: the patient knows whether this has been done or not. Or consider something as simple as flossing; it is hard to see how one could substitute a placebo.

Thus some treatments cannot be tested because it would be wrong to test them (e.g., smoking); some cannot because it would be impractical to do so (again, smoking); and some cannot be tested even in principle (e.g., exercise).

Taken together, publication bias, financial bias, and pill bias significantly skew the kinds of treatment that EBM is likely to prescribe. Chemically novel pharmaceutical treatments for persons who are affluent or are covered by affluent national health schemes are more likely to be prescribed, because evidence from RCTs or systematic reviews of RCTs is more likely to be available for them. Treatments that are not new, or otherwise not patentable, are less likely to be prescribed. Treatments that it is unethical, difficult, or impossible to test with an RCT will be less likely to be prescribed, and where available, pharmaceutical treatments will typically be prescribed because better evidence will be available. An obesity pill, if one could be found, would be supported by much "stronger" evidence than could ever support exercise; yet it is very hard to imagine an obesity pill with better overall health benefits than a well-constructed exercise program. Antidepressants will always be supported by "stronger" evidence than cognitive-behavioral therapy. Many esoteric, fringe, and alternative therapies obviously suffer, but so does surgery, often supposed to be the pinnacle of modern medical achievement.

What significance do the imperfections of RCTs, and of the biomedical research system more generally, have for EBM? The reasonable conclusion of these criticisms, in my view, is as follows. These problems show that any injunction that clinicians rely in a quasi-automated way on published research—whether from RCTs or something else—is going to be problematic. If EBM is taken as urging a very strict reliance on published literature, then its effect will be to pass these problems on to clinical practice. If, on the other hand, EBM is seen and practiced as a more general kind of encouragement to consider published research in clinical decision-making, then these problems are not pressing. However, the problems that EBM was supposed to overcome will re-emerge. Doctors will once again have room to express doubts about the credibility of research, or opinions that go beyond it; and their own biases will operate once again. The fact that the biomedical research system is a rather leaky vessel is thus a significant difficulty for EBM, and one that renders its prescription and the consequences thereof much less clear than first appearances suggest.

5.5 THE PROBLEM OF TRANSPORTABILITY

EBM seeks to combat bias, but it also seeks to promote specificity. As already mentioned, a laboratory experiment will typically be *less* subject to bias than even an excellent RCT. The reason EBM places evidence from the laboratory at the bottom of its pyramid is not that it is subject to bias, but rather that it is (typically) not specific to the effect of the treatment in question. Evidence that a certain substance inhibits growth of cancer cells in vitro might lead us to infer that, if the substance is placed in a pill and swallowed, cancer sufferers will experience remission. But bitter experience shows that such inferences are not reliable. The body is complex and reacts in unpredictable ways. EBM espouses direct tests of the effectiveness of the proposed treatment in the real world—tests involving real people swallowing real pills. It places little reliance on chains of reasoning from evidence gained in the rarefied setting of the laboratory.

EBM recommends reliance on evidence that bears as directly as possible on the treatment whose outcome you are trying to predict, and thus prefers evidence that comes from the "real world" as far as possible. This reduces the number of inferential steps that one must take to conclude that a given treatment is likely to have a certain effect on a certain patient.

EBM is subject to a corresponding criticism, namely that its methodological prescriptions in fact fail to achieve this. In outline, the criticism is that the RCTs are rather artificial, and not much like real life; and thus that the very same problem arises for using the results of RCTs to predict the effectiveness of a treatment in a particular patient as for reasoning from laboratory science.

In a clinical context, the notion of "best evidence" (or even "good evidence") is thus potentially misleading. Evidence from an RCT may be the best protected from bias, but the circumstances of the RCT may not resemble those of the clinic very closely. In general, there is a tradeoff between the admissibility criteria for an RCT and the number of people on whom the result has a very direct bearing. Thus, for example, if one excludes from a study of antidepressants those who are above 65 and below 25, then the results bear less directly on the young and old. If one excludes pregnant women, as is common, then the results do not bear directly on pregnant women. If one excludes persons who are taking medication for epilepsy, blood pressure, and diabetes, the group of people to whom the results directly apply reduces further still. And if one confines one's study to persons of similar background, race, culture, socioeconomic status, diet, exercise habits, adiposity, and so forth, then the problem compounds.

The trouble is that trials often do implement such exclusions, in an effort to improve the methodological credentials of the trial (or, more cynically, to find a positive result). There are at least two reasons for this: confounding and interaction. First, the more uncontrolled factors that potentially confound a finding, the greater the chance that one of them does in fact interfere with the outcome. In theory, randomization should take care of this; but as already noted, the odds are 1 in 20, and Russian roulette would still be an exciting party game even at these odds. In particular, if adverse events occur in a trial, there is a serious chance it will need to be stopped on a precautionary basis, without waiting to establish causality. So including groups who are already at a higher risk of death or other adverse events is a risk to the completion of the trial itself. Second, some of these things may interact with the treatment being tested, in ways that cancel out or amplify the signal that one is trying to detect. Some drugs may affect the young and old differently; some may interact with other medication; and so forth.

In the clinical context, however, these exclusions become problematic, because the standard exclusions are often *typical* of the clinical setting. Depression, for example, is particularly common in youth and old age, among the obese, and among those suffering chronic illnesses; many of these will take some other chronic medication. Pregnancy poses a particular problem, because very few tests include pregnant women for fear of harming the fetus. But women who are on chronic medication fall pregnant, and are then often faced with a choice between stopping medication and continuing with unknown risk to the child. Surprisingly little research is done even on the safety of drugs during pregnancy, let alone on their efficacy. Epileptics on medication to prevent seizures face a choice: risk harming the fetus (and even themselves) by continuing their medication in pregnancy, or stop the medication and increase the risk of seizure, which can also harm the fetus or cause miscarriage. The choice is really a guess, since both the safety and the efficacy of the drug during pregnancy is almost never known.

Nancy Cartwright has been the most vociferous philosophical critic of EBM and its policy cousin, Evidence-Based Policy (Cartwright 2007a; Cartwright 2007b; Cartwright 2010; Cartwright 2011; Cartwright and Hardie 2012). She famously distinguishes two questions: "Does it work somewhere?" and "Will it work for you?" The irony is that EBM sets out to answer the second question ("Will it work for you?"). But there is often a tradeoff between methodological impeccability and real-world applicability in RCTs. This raises a serious problem for EBM: the better the RCT, the longer the chain of inferences will typically be between the study and the particular clinical situation.

On one way of looking at it, EBM falls into the same kind of trap that it accuses traditional reasoning from biological, chemical, and anatomical knowledge of falling into. It emphasizes the rigor of the local investigation and neglects the question of whether what is discovered locally will apply in practice. For the traditional medical scientist the basis of this reasoning is the action of a substance on a model species, culture in a petri dish, etc. For the evidence-based researcher, it is reasoning from the action of a treatment in a studied population. EBM points out sharply that no matter how good the lab science, things in the real world don't always go as expected. Cartwright and others who have pressed the difficulties of transportability on EBM proponents are giving them a taste of their own medicine, in this sense, because they make exactly the same point about RCTs: no matter how good they are, things in the real world—the world beyond the studied population—don't always go as expected. Put strongly, the point against EBM here is that it has failed to solve its own problem, because it has failed to provide or even to improve upon methods for determining the probable effectiveness of treatments in clinical settings.

The underlying philosophical point is that it is a fundamental error to think that the quality of evidence can be assessed *solely* by reference to the design and implementation of the study that produced it. The problem is that the quality of evidence depends on what the evidence is *for*. In a situation where we can reasonably assume that effectiveness of a treatment will not differ between populations, we can reasonably rely on published evidence in making clinical decisions. Unfortunately, many situations are not like that. Effectiveness is not an intrinsic property of a treatment, like molecular weight. It is more like a secondary property, such as color, whose perception depends upon the presence of a visual system of a certain kind. Effectiveness of a treatment likewise depends upon the presence of a suitable set of circumstances.

The natural reply is that, nonetheless, an emphasis on testing treatments is surely useful. This is true, and I will return to this below when I consider the real and lasting contributions of EBM to medicine. But EBM sets out to expose the fact that treatments are often prescribed or recommended when there is no good reason to think that they will work. It is fatally damaging to this project if the proposed solution is likewise open to the criticism that it warrants treatments for which there is no good evidence of probable effectiveness. In my view, this point shows that EBM is a failure at the conceptual level, whatever practical benefits it might rightfully claim.

5.6 WHAT'S GOOD ABOUT EBM?

I have canvassed criticisms for EBM, and concluded that it is both a practical and conceptual failure. Is it really that bad? What about the initial appeal of the idea that we should try to base our decisions on best evidence? What about the social project that I alluded to, of upending medical hierarchies? Has the effect on medicine been positive?

One of the most unfortunate things about EBM has been the tone and manner in which it has been advanced. The central claims of EBM are indeed too strong, and their initial statement was an overstatement, subject to many successive modifications and qualifications. Some of the leading lights tell stories of publicly humiliating others whom they deemed to be irrational, authoritarian, or otherwise deserving. It is evident from these autobiographical accounts that there was some aggression in the manner in which the movement was promoted, at least at times. The response on the part of those individuals would no doubt be that this was a necessity, because the medical profession would not otherwise have listened; and that, after all is said and done, EBM changed the profession for the better, whatever the quibbling philosophers might say about its conceptual shortcomings.

Has EBM improved medicine? This question is hard to answer. On the one hand, EBM led to the Cochrane Collaboration, and this is very useful. Clinicians and patients alike can access evidence from published studies more readily than if the Cochrane Collaboration did not exist; and indeed so can anyone with access to the internet. Cochrane is here to stay, and that is very probably a good thing. Moreover, clinicians and ordinary people are arguably much more *likely* to access medical research as a consequence of EBM exhortations. This demand has arguably led to the faster and better development of other online tools such as PubMed.

On the other hand, some of these developments would almost certainly have occurred without EBM. The rise of the internet has first enabled and then generated demand for similar developments across the field of human interest. It is not clear how much of the credit EBM as a movement can claim for the extent to which clinicians now consult published population-level research when making clinical decisions, given that it was simultaneous with the rise of the internet, which is a necessary tool for consulting the Cochrane database. Indeed, the style of EBM may even have put some physicians off. It is also questionable, as we have seen at length, that EBM gives the best advice about what to do with the evidence that it exhorts patients and practitioners alike to locate.

The evidence hierarchy has shaped biomedical research, and, given the problems with that hierarchy, it is hard to see that as a helpful contribution. In addition, the helpfulness of databases and online resources might actually be *greater* if the evidence pyramid had never been proposed. EBM also encourages a rather strange view of the source of medical knowledge, most of which is derived from before RCTs were even invented. Our most effective interventions have not been subject to RCTs; EBM has not in fact yielded our most effective interventions. The point is the subject of a famous lampoon, a paper recommending a randomized crossover trial of the hitherto-untested parachute, with the most vocal proponents of EBM as its participants (Smith and Pell 2003).

It is hard to get enough historical or analytic distance from EBM to answer this sort of question. However, EBM's most plausible claim to have positively influenced medicine is probably that it has created a space in medicine for the idea that a clinician can appeal directly to published research in clinical decision-making, and that it has encouraged much larger numbers of clinicians than otherwise would to do so. This probably does amount to a difference between the way clinical decision-making occurs now from how it would have occurred without EBM, and it is probably a positive difference.

5.7 CONCLUSION

I have spoken about EBM in the past tense. This is because I consider it to be a movement within medicine, and not a type of medicine; and I consider that movement to have now come to a halt. Parts of EBM have been incorporated into Mainstream Medicine. Those parts include both ideas and people: those who were once the young Turks are now establishment figures.

I have also been primarily critical of EBM. In part this reflects my own personal reaction: I struggle with what seems to me to be a hypocritical espousal of doctrine and skepticism in the same breath, with the apparent lack of awareness of how medicine got to the point it has got to, with the unrealistic and potentially dangerous (I believe) espousal of formulaic and procedural decision-making processes in place of judgment that cannot be explicitly justified. Others, however, see it differently: some see the simple messaging as a necessary and effective means for getting a message across; some see historical awareness as an academic luxury independent of the practical needs of clinical medicine; and some see the proceduralist emphasis as a necessary and helpful corrective to the lazy guess or uncritical reliance on authority.

In this chapter I have left out an important component of the clinical picture, and indeed of EBM—namely, the patient. EBM is intended to be "on the patient's side," giving the patient tools to catch out bad doctors, and the courage to ask doctors to justify their prescriptions. However, EBM suffers from another tension here, because the *kind* of evidence that EBM prioritizes is population evidence. A clinician can of course take individual features of a patient into account, but ideally such deviations should also have an evidence base. The thrust of EBM really militates against doctors deviating on the basis of a feeling that a given patient just won't get along so well with this drug, or anything of that nature. This is part of the proceduralist character of EBM decision-making, and it is what is meant to protect against unconscious or conscious biases. Yet patients may well feel very confident that they are atypical.

This is not exclusively a problem for EBM, of course: it is a problem for Mainstream Medicine in general. For this reason we will consider it in the next chapter, in the context of what has become known as patient- or person-centered medicine.

Chapter 6

Medical Nihilism

6.1 IS MEDICINE ANY GOOD AT ALL?

A thread running through the book thus far is that medicine is not all it's cracked up to be. The history of medicine is not a history of dramatic curative success, and contemporary claims to success are accompanied by areas of continued curative frustration. EBM is more optimistic, but it starts from the idea that claims of effectiveness are unwarranted unless they are "evidence-based," and EBM restricts the meaning of that term in a way that excludes the basis of many medical treatments in past and present use. In the present chapter, I want to finally confront this theme directly. I want to consider the idea that there is something deeply wrong with medicine, and that it is less beneficial than commonly thought, not beneficial, or harmful. This is *medical nihilism*, in the broadest sense of that term.

In everyday usage, Nihilism is the view that life is meaningless and worthless. This view is not mere skepticism about the meaning of life—it is not the mere doubting that life has meaning or worth. It is the conclusion that it does not. It has an emotional component: despair and disillusionment. The term "Nihilism" has these connotations and its application to medicine yields the view that medicine is worthless, accompanied by an emotional response of disillusionment, despair, and despondency. It is not mere skepticism, or doubt, about medicine: it is not merely the request for more information in order to be able to make up one's mind. Rather, it is the result of making up one's mind, and forming a belief about medicine, accompanied by a negative emotional response.

In the 19th century, Medical Nihilism was a common stance in Western Medicine. It arose from the continued impotence of medicine to cure disease, despite advances in understanding. A number of people in the profession became disillusioned. Probably the most famous expression was by the then Dean of the Harvard Medical School:

> Throw out opium, throw out a few specifics . . . throw out wine . . . and the vapors which produce the miracle of anaesthesia, and I firmly believe that if the whole materia medica, as now used, could be sunk to the bottom of the sea, it would be all the better for mankind,—and all the worse for the fishes.
>
> (Porter 1997, 680 [quoting Holmes])

Porter emphasizes that such sentiments could be expressed in pauper hospitals, or high-minded settings (such as Harvard), but not in common private practice, where they would have been "suicidal" (Porter 1997, 681). Nonetheless, it is clear that a significant number of thoughtful medical persons held nihilistic views about, and attitude toward, medicine. They did not merely doubt medicine: like Holmes, they "firmly believed" that, with a hodgepodge list of exceptions, medical treatments did too little good and too much harm. Skepticism about medicine is uncertainty. Medical Nihilism, on the other hand, is a strong belief in the impotence of medicine (setting aside that hodgepodge of exceptions). That is why Stegenga calls it "a radical position about medicine" (Stegenga 2018, 11).

This is not the 19th century, however. A natural reaction to Contemporary Medical Nihilism is to ask: why the hostility? Surely medicine is not that bad? Isn't at least contemporary Mainstream Medicine, properly practiced, a life-saver and changer, and a huge force for good in the world? (And if you favor some other tradition or medical discipline, you can ask the same question about that one.) Would you not appreciate this if you suffered from any one of the countless afflictions that medicine can deal with successfully—appendicitis, meningitis, cataracts, and a long-continuing list? Isn't nihilism just carping from the sidelines? Is it even, dare I say it, jealousy on the part of over-educated and neurotic commentators, who, many years back, chose a less useful path in life—chose to study philosophy or history or socio-anthro-you-name-it-ology, because it gave them more time and material for posing with cigarettes and long coats, and did not require them to do any mathematics?

This sort of ad hominem response is, of course, unsatisfactory from a logical point of view. But it also fails to apply to the most prominent individuals associated with medical nihilism, who, as I have indicated, are often senior medical persons. It is true that the two nihilists on whom I will focus on this chapter are a historian and a philosopher. However, the historian, David Wootton, confines his nihilism to the past; he is very positive about contemporary Western medicine and particularly EBM. Jacob Stegenga, the philosopher, is nihilist about contemporary medicine (including EBM), but his early career was in the biomedical sciences, and so again, the ad hominem would not apply even if it were a good argument.

It is interesting that Medical Nihilists are often medical insiders. As Stegenga puts it:

> The work written by physicians, epidemiologists, and science journalists supporting medical nihilism is vast. These thinkers are not cranky outsiders, but rather are among the most prominent and respected physicians and epidemiologists in the world.
>
> (Stegenga 2018, 13)

The fact that medicine itself has produced a string of distinguished nihilists suggests that we ought to take Medical Nihilism seriously, because it suggests that people in the know detect something wrong.

6.2 CLARIFYING MEDICAL NIHILISM

Medical Nihilism as I have defined it is broad, and it is useful to clarify the possible ways in which medicine might be deemed to fail.

First, we can distinguish between failure with respect to different *metrics of assessment*. We could have Curative Nihilism, the failure of medicine to cure; we could have Therapeutic Nihilism, the failure of medicine to offer effective interventions more generally, even those falling short of cure (recalling the distinction between cure and therapy in Chapter 2); and each of these may or may not suffice for full-blown Medical Nihilism, depending on how we assess the success of medicine as a whole. If the metric of assessment if complete cure, then Curative Nihilism will warrant Medical Nihilism; if it is effective intervention more broadly, then Therapeutic Nihilism will warrant medical nihilism; and if medicine can be deemed successful even when it does not offer effective intervention, then Medical Nihilism will not be warranted by either Curative or Therapeutic Nihilism, and will require further justification beyond showing that medical interventions are ineffective. Obviously, I am particularly sensitive to these distinctions because they relate to my own position, and, in my mind, offer the most plausible way out of Medical Nihilism.

Second, we can distinguish between failure in different *domains*. One could be a *Universal Nihilist*, contending that every aspect of medicine is worthless or fails to reach its proper goals. But I have not encountered any Universal Nihilists. Most Medical Nihilists exempt certain areas of medicine from their mistrust. Oliver Wendell Holmes exempted "the miracle of anesthesia," along with a few "specifics" (which, recall, were medicines that are effective for specific diseases, but are not part of any larger therapeutic system, and do not seem to offer any opening for extension or development). As we shall see below, Wootton exempts contemporary Western medicine since the advent of EBM and the rise to prominence of clinical trials, while Stegenga goes the other way, doubting modern treatments that are typically supported by evidence from clinical trials but exempting a handful of "magic bullet" interventions mostly developed 70 to 90 years ago. I will call the view that historical medicine is bad *Historical Nihilism*, and when it is coupled with the assertion that contemporary medicine has finally escaped the problems of historical medicine, *Whiggish Nihilism* (which is provocative, but accurate). I will call nihilism about contemporary medicine *Contemporary Nihilism*. Wootton and Stegenga are both Historical Nihilists; in addition, Wootton is a Whiggish Nihilist, while Stegenga is a Contemporary Nihilist.

Third, we can distinguish different *strengths* of nihilism. A Weak Nihilism would say that medicine is less beneficial than one might suppose; a Moderate Nihilism would say that it is not beneficial at all; and Strong Nihilism asserts that medicine does more harm than good.

These three clarifications will help us make sense of the ideas of some actual Medical Nihilists. But just before we get there, it is worth noting that even with distinctions and clarifications in hand, Medical Nihilism remains quite a difficult position to get clear on. When assessing Medical Nihilism, the plausibility of the position often seems to come down to matters of degree, and to unsatisfactorily subjective assessments of the relative weights of good and bad "bits" of medicine (there being no serious Universal Nihilists). Do we focus on the wonders of penicillin, or the disappointments of antivirals? Do we consider appendectomy to be a marvelous success, or a brutality to which we have grown accustomed, the substitution of a deliberate harm from which the body is more likely to recover for an accidental ill from which it is less likely to recover? Do we assess statins by their modest contribution to individual health, reducing the relevant risk by around 2%, or the large number of lives that a 1% to 2% risk reduction saves when it concerns an exposure that is very prevalent in a population?

This relates to the key question of how one is to assess the success of medicine. Later in this chapter I will argue that one reason Medical Nihilism fails is that it tends to assess the success of medicine solely by its success at cure: in other words, it collapses Medical Nihilism in general into Therapeutic Nihilism, which is despair about cure. In doing so, it wrongly neglects the success of medicine at pursuing its core business, which is not cure, but inquiry. Failure to reach the goal of cure does not warrant Nihilism so long as that business is growing, and growing in the direction of cure. Even Therapeutic Nihilism is not warranted in these circumstances, I will suggest. It is only where our efforts to understand draw a continual blank that we ought to really worry.

6.3 WOOTTON'S WHIGGISH NIHILISM

David Wootton is a historian, and his Nihilism concerns medicine's past. His is a Strong, Whiggish, Therapeutic Nihilism. That is, he believes that in the past, medicine did more harm than good, and that its failure to benefit is to be measured against failure to offer therapeutic interventions even falling short of cure; and he believes that contemporary medicine is contrastingly successful, and that Nihilism is not warranted about it.

Wootton's argument is historical. He writes a history of Western medicine in which the harm inflicted by doctors on their patients is the narrative theme. His history paints a gory and disconcerting picture. However, Wootton is primarily skeptical of historical medicine. He holds that, while there is some bad medicine around, we now have the tools to root it out, and have made significant steps toward doing so.

The correctness of Wootton's Nihilism depends on three factors. The first is the success of historical Western medicine at administering cures. We can agree with him that it was not very successful, and often harmful, even if more even-handed historians (such as Porter and Bynum) do not paint such a dramatic picture, and in particular not a picture that paints patients as victims of knowingly harmful doctors. Historical *Curative* Nihilism is hard to deny, even if one sees other values in medicine besides cure (as I do). The second is the success of contemporary Mainstream Medicine at administering cures. Wootton maintains that Mainstream Medicine is uniquely curatively successful, a claim that many will find plausible, although not all, as we will see in the next section. The third is a claim about *why* contemporary Mainstream Medicine is successful, and has safeguarded itself against future error.

Wootton holds that the rise of EBM justifies confidence in contemporary medicine. This stance sits uneasily with the date on which he claims modern medicine began. That date, 1865, is over a century before the advent of EBM, and not much less than a century before Austin Bradford Hill and others devised the RCT study design. His faith in EBM is also anachronistic from a philosophical point of view. As we have seen in Chapter 5, EBM faces some serious problems. My own view is that its prescription for medicine is neither practically nor theoretically desirable. Contemporary research methods, including RCTs, are very far from perfect, and clearly do not safeguard us from error.

If medicine is an inquiry with the purpose of cure, as I maintain, then Wootton's Whiggish Nihilism is flawed for another reason. It draws a sharp line between past and present, based solely on curative effectiveness. But where do these cures come from? Why did William Lister decide to employ antiseptic procedures in 1865? Surely the process of inquiry that led him to do this must count as part of the medicine he practiced as much as the actual act of employing antiseptic procedures itself. Cures do not pop out of medical heads like lightbulbs going on. One can acknowledge that cures became dramatically more available at a certain point, without agreeing that medicine before that was worthless. Indeed, it was from the earlier medical tradition that the later one emerged. Even completely incorrect medical notions might serve a purpose in the grand scheme of things. One has to start somewhere, and most probably the first hypothesis will be wrong. Eliminating falsity has value, even if not as much value as obtaining truth.

There are really two aspects to my objection, connected though they are. One is that, if the core business of medicine is inquiry, it can succeed at this even when not attaining the goal of cure, because one can nonetheless make progress with the inquiry, even while not attaining truth in the inquiry. This is, in turn, because progress is possible even before one gets the right and final answer. The second is that cures come from the business of inquiry. They may come much slower than we would like, but that only makes my point more pressing. It is weird to celebrate cures while condemning the process that yielded them. Contemporary public discourse hails breakthroughs in medical science for having curative promise, long before practical applications are devised (indeed, many never are). And so it is weird to insist on a change of epoch at the point of a successful curative breakthrough, such as the implementation of antiseptic procedures, and designate that as the beginning of the end of bad medicine. Any such beginning must itself have begun earlier, with the possible exception of a situation where a cure is handed down from on high.

Wootton's Whiggish Nihilism is thus not compelling. It seems to pin its faith in modern medicine on EBM, which is naïve. It does not explain its attitude to medicine that occurred between 1865 and about 1995 (for sake of argument), given that medicine prior to the 1990s was very rarely evidence-based, and yet the date given for the start of modern medicine is 1865. It does not explain its confidence in those parts of contemporary medicine that are not evidence-based. And it does not leave room for the continuous character of medicine. Just as Newton was both alchemist and scientist, Lister was a practitioner of medicine both before and after the watershed year 1865. "Good" medicine arose from "bad," just as science arose from alchemy. It is as misleading and melodramatic to call medicine before 1865 "bad medicine" as it would to call alchemy "bad science." Sell books it may, but it does little to advance our understanding of medicine.

More fundamentally (and forgivably), Wootton's view misconstrues medicine. It adheres to something like the (intuitively appealing) Curative Thesis, which asserts that cure is both the goal and the core business of medicine. However, I have sought to refute the Curative Thesis. Instead, I have advanced and defended the Inquiry Thesis, which asserts that the core business of medicine is not cure, but inquiry with the purpose of cure. One can accept that medicine fails to reach its goal of cure without giving up hope, because one can believe that its core business is progressing. And indeed, it is only through progress in this core business that we can explain the curative breakthroughs that Wootton celebrates. It will not do to call medicine "bad" merely because it does not cure, when it is engaging in an inquiry that eventually produces a cure. One must do more than show what Wootton has shown, which is a litany of curative failure. One must also show failure of progress in inquiry. This Wootton has not done, and so his work does not justify Whiggish Nihilism.

6.4 STEGENGA'S CONTEMPORARY NIHILISM

Jacob Stegenga argues for Contemporary Medical Nihilism. This is not merely the view that one should be skeptical, cautious, wary, and so forth when countenancing claims of medical effectiveness. It is more than that. Stegenga writes:

> Medical nihilism is not merely a tough skepticism espousing low confidence about this or that particular medical intervention. Rather, medical nihilism is a more general stance.
>
> (Stegenga 2018, 10)

This more general stance is that "our confidence in the effectiveness of medical interventions ought to be low" (Stegenga 2018, 11).

The reason for low confidence is not historical, but contemporary. Stegenga does mention the history of curative failure but is clear that this is not the basis of his argument. He argues that, when we consider some evidence that a given intervention is effective, we should bear in mind a number of factors: "the frequency of failed medical interventions, the extent of misleading and discordant evidence in medical research, the sketchy theoretical framework on which many medical interventions are based, and the malleability of even the very best empirical methods" (Stegenga 2018, 10). He argues that, when we take these things into account, we ought to assign a low probability to the claim that an intervention is effective, even after considering apparently strong evidence in favor of its effectiveness.

Stegenga's view is supported by the following "Master Argument":

> on average, we ought to assign a low prior probability to a hypothesis that a medical intervention will be effective [due to factors such as those listed above]; . . . when presented with the

evidence for a hypothesis we ought to have a low estimation of the likelihood of that evidence; and similarly, . . . we ought to have a high prior probability of that evidence.

(Stegenga 2018, 14)

This is an argument employing Bayes Theorem, a mathematical theorem whose deductive validity is not in question. The theorem is:

$$P(H|E) = \frac{P(H)P(E|H)}{P(E)}$$

The "prior probability" of H is the probability that we, subjectively, assign to a hypothesis when we first encounter it, before considering any evidence. Stegenga says this should be low: we should have an initially skeptical attitude to claims of effectiveness of medical interventions, before we have considered any evidence. This is because of factors such as the frequency of failed interventions.

Stegenga also says that the prior probability of the evidence, $P(E)$, should be high. This is justified by the malleability of research methods. In particular, various kinds of bias mean that, as well as having low confidence that the resulting hypothesis is true, we should also have a rather high confidence that the evidence will favor it. In general, we should *expect* to encounter evidence that a drug is effective, because we know that financial interests, publication bias, and so forth operate generally, and because we know that detecting their operation in any particular case is tricky. Bayes Theorem has the effect that, when we encounter unsurprising evidence (with a high prior probability), it does not add much to our reason to believe a hypothesis. The intuitive idea behind this is that if you expect something to be the case "anyway," so to speak, then observing that it is the case does not give you much further reason to believe any particular hypothesis. If I claim to be a prophet and predict that the sun will set this evening, you will not be convinced by the subsequent setting of the sun, since you were expecting that. If I predict that the sun will bounce along the horizon like a rubber ball, however, and to your surprise it does so, you will give me more credence.

Stegenga also claims that likelihood of E is low. This is a technical use of "likelihood," which is not synonymous with the generic term "probability." In Bayesian parlance, "likelihood" is one probability in particular: the probability that one would observe the evidence in question if the hypothesis in question were true. Suppose you are drawing balls from an urn, and your hypothesis is that all the balls in the urn are red. You draw a ball and it is red: thus E is "This ball is red." The likelihood of the evidence is then 1, since if all the balls are red, then the probability of drawing a red ball is 1 (or 100%). But if your hypothesis is that half are red and half are not red, then the likelihood of the evidence is 0.5, since if that were the case you would have a 50% chance of (randomly) drawing a red ball.

Stegenga's claim that the likelihood is low highlights a problem in his presentation. Strictly speaking, if H is just the hypothesis that the intervention is effective, then it cannot be right to say that the probability of finding E given H is low. E is evidence of low effectiveness, but that is still effectiveness, and H just says that the intervention is effective. So, H renders E very probable indeed, along with any other evidence that shows effectiveness. However, Stegenga's notion of effectiveness is richer than that. Elsewhere in the book he indicates that he has a higher and more sophisticated standard for effectiveness than the mere presence of some effect (Stegenga 2018, 40–46). In other words, H is really something like "The intervention is highly effective," or "The intervention is more than merely effective," or "The intervention meets Stegenga's criteria for effective intervention." Stegenga's thought is that, on such an interpretation of H, E becomes improbable, because E is the kind of evidence we typically get: evidence of low effectiveness. In short: the hypothesis that the intervention is highly effective makes evidence of mild effectiveness improbable. This response is not adequate, because Stegenga also thinks the probability of evidence of effectiveness in this sense is high—he believes P(E) is high. Only if the sense of "effectiveness" changes from its appearance in E and H can this be a defence of the low likelihood claim. However let us pass on this technical point for the sake of argument.

If $P(H)$ and $P(E|H)$ are low and $P(E)$ is high, then $P(H|E)$ is the result of dividing a small number by a large one, meaning that it will be low. Thus the confidence that we should have in effectiveness ought to be low, even after considering a piece of evidence for effectiveness. And this, Stegenga contends, is the essence of Medical Nihilism.

Stegenga's exposé of medical research is powerful and worthwhile. The question I want to focus on, however, is what attitude it justifies, and in particular, whether it justifies Medical Nihilism.

6.5 KINDS OF EVIDENCE

There are a number of questions and objections one could raise to Stegenga's argument. I want to focus on a question that will arise for any other defense of a similar Contemporary Medical Nihilism. This question asks: what is the relevance of background information about the characteristics and performance of biomedical research? Looked at one way, it is obvious that these are relevant to any decision about how much confidence to place in the outcome of such research. Failure to do so would be to commit an error of reasoning, known as the Base Rate Fallacy. Looked at another way, there is a risk of over-generalization, of tarring all with the same brush. Can we really generalize about all biomedical research from the study of a few cases?

Stegenga partly anticipates such a concern:

> Perhaps the most vociferous objection will be that medical ni-
> hilism is a general and empirical thesis, but I have merely selected
> a handful of examples that appear to be favorable to the thesis.
>
> (Stegenga 2018, 14)

At points he seems to respond to this argument by insisting that the
examples are in fact typical, which amounts to saying that the generalization is warranted. For example:

> One might charge me with cherry-picking examples of medical
> interventions that are barely effective. However, these examples
> are among the most widely used drugs today, and such examples
> are ubiquitous—it takes little effort to expand lists of such
> examples.
>
> (Stegenga 2018, 216)

However, this is not a strong response, as he elsewhere acknowledges
(Stegenga 2018, 14), because it does not answer the objection. That
objection is that inferring so much from such a small sample is infer-
entially weak. That weakness is not fixed by simply denying it.

Stegenga does provide a much stronger argument for generalizability. He emphasizes that his argument does not have the structure
of a simple inductive generalization: it does not consist in exhibiting a
few cases and then inferring something about the whole field. Rather,
it consists in identifying the methodological, social, financial, and
other factors that *give rise to* the problems in these cases—those that
explain it—and, in effect, generalizing these. It is thus an inference
to the best explanation, or abductive inference, rather than an enu-
merative induction or a generalization. The best explanation of the
features that are exhibited in the examples is the operation of various

factors, and this explanation supports a generalization across much of medical research science. This generalization is well supported: it enjoys some direct independent evidential support (such as evidence showing the operation of financial bias across the industry) and support from the sheer implausibility of supposing that a bias that was detected or revealed in one case was an isolated instance. Mostly, biases and errors are concealed, either deliberately or by our own blindness to them, so it is much more likely that we under-detect them than that we detect them only in the cases where they occur. Thus all Stegenga needs for his argument is the contention that the cases he focuses on arise from widespread features of biomedical research and its context. And this is indeed very plausible, given the biases and other features that he picks and his explanations for them.

However, the problem of overgeneralization is deeper than this, and two aspects of it survive Stegenga's response. The first concerns the *kind* of medical evidence that his critique focuses on, which I deal with in this section, and the second concerns the *relevance* of background information on individual cases, which I deal with in the next.

The trouble is that Stegenga's methodological critique only concerns those interventions whose effectiveness is supported by those methods that are subject to the critique. It does not follow for others. This is obvious, and indeed a point to which Stegenga himself implicitly appeals; he accepts that there are "wonder drugs," mostly devised before clinical trials. Stegenga also notes that he is primarily confining himself to pharmaceutical interventions, which is true, for surgery cannot be subjected to trials with the same ease as pills, and in fact generally isn't; moreover, surgery is not subject to the same commercial interests as pharmaceuticals. If Stegenga's premises concern the entirety of medicine, then his conclusion follows logically; but he himself acknowledges that they do not, and therefore his conclusion either does not follow or, more charitably, does not concern the whole of medicine. To the extent that it depends on a critique of medical research methods, Stegenga's Medical Nihilism is not Universal Nihilism; it is Nihilism about pharmaceutical interventions whose primary evidence of effectiveness comes from clinical trials. To call this "Medical Nihilism" is a little overstated, since there is much more to medicine than those interventions, a point that Stegenga recognizes in the context of both "magic bullets" and what he calls "gentle medicine," which is his recommendation for medicine, a harking back to Hippocratic modesty about curative power and personal concern for the individual patient. Stegenga makes clear that his Nihilism is really directed at the products of the contemporary pharmaceutical industry, and not at vaccines, antibiotics, surgery, and other areas. However, he also points out that an inordinate amount of contemporary medicine—and its money—is focused on pharmaceuticals of the kind he criticizes; there is relatively little work or money going into new antibiotic research, in contrast. Nonetheless, the term Medical Nihilism might reasonably be criticized as rather broader than the underlying critique.

One response to this criticism would be to assert that the best evidence comes from clinical trials, or other formal trials, and not from any supposed "obviousness" of effectiveness in the clinical setting, nor from reasoning from anatomical and biological knowledge, nor any other source. Such a response is not one that Stegenga himself would want to make, since it amounts to endorsing a kind of evidence hierarchy, and Stegenga criticizes evidence hierarchies in depth. Nor is it an easy path for anyone who is not Stegenga to follow, since they would have to deal with Stegenga's criticisms, which are compelling.

The problem, at its core, is that any critique based on the methodological factors, construed generously so as to include the influence of the social, is going to justify Nihilism only insofar as medical research uses these methods and is subject to these social factors. While it is plausible that some forces, notably greed, operate very widely, efforts to demonstrate the mechanisms by which such forces can operate—such as Stegenga's demonstration of the malleability of research methods—inevitably tend toward the specific. The core of the problem is that medical knowledge derives from various sources, and the majority of it does not in fact derive from biomedical research of the kind that Stegenga criticizes. Much still derives from anatomical and biological knowledge, and from handed-down practices. And, as is often noted, even the application of evidence-based recommendations to an individual often requires a degree of reasoning of this nature, to determine whether specific features of the individual might adversely or positively affect the prospects of successful treatment.

For example, administration of adrenaline during cardiac arrest is standard practice, but it is not evidence-based. This lack of evidence is hardly to be celebrated, and may explain the American Heart Association's tentative guideline:

Standard-dose epinephrine (1 mg every 3 to 5 minutes) may be reasonable for patients in cardiac arrest. (Class IIb, LOE B-R).
(https://eccguidelines.heart.org/index.php/circulation/cpr-ecc-guidelines-2/part-7-adult-advanced-cardiovascular-life-support/ , Sec. 5.3.3.1.2)

There are attempts to test the effectiveness of this common intervention and identify an evidence base, but evidently these attempts lag behind the use of the actual procedure.[1] This is by no means comforting: it is not something to be celebrated. It might well be a cause for some skepticism, or doubt, about the effectiveness of this intervention, on the basis that we know how easily reasoning from biological and anatomical knowledge combined with an acceptance of what has been done before can go wrong, and we might conclude that

1. I am grateful to Jeremy Simon for helping me substantially with this point.

a sufficient quantity of evidence has not yet been gathered for us to be confident that it is effective. However, we do not have any reason to think that it is probably *not* effective. It may be more likely effective than not, but we might want more evidence nonetheless because this balance is based on such a paucity on both sides. One cannot apply Stegenga's critique to this virtually standard procedure, since that procedure arises from reasoning about the biochemistry and anatomy of the human body. Thus a conclusion that the intervention is probably ineffective, the Nihilistic view, is unsupported.

Much medicine derives from such reasoning, and a very great deal remains without significant testing in trials. We might be skeptical about such medicine, but skepticism is not Nihilism: the firm belief of worthlessness. Nihilism requires a positive justification, such as Stegenga's arguments seek to supply, and those arguments do not apply where medical knowledge derives from sources other than those he criticizes.

6.6 THE BLUE BUS PROBLEM

The second aspect of the generalizability problem for Nihilism concerns the relevance of facts about the whole of medical research for forming opinions about particular interventions. There is a famous legal problem known variously as the Blue Bus Problem or the Gatecrasher Paradox, among other names. In the Blue Bus version, a plaintiff is driving at night when a bus hits the car. The plaintiff did not recall the bus's color, but 75% of buses in town (or on that route, at least) are run by the Blue Bus Company. In civil liability, the standard of proof is the balance of probabilities. More likely than not, the bus was a Blue Bus. Should the Company be held liable? The numbers seem to indicate that it should, but courts balk at such a suggestion, demanding that evidence about the case in question be brought to bear.

However, even if such evidence is brought to bear, the question then becomes whether background information remains relevant. Does the Blue Bus Company have a heavier burden to discharge because it owns most buses in town? Similarly, if 1,000 people come to a baseball game but only 499 pay for their tickets, can the stadium sue any randomly selected person for the price of a ticket provided she can prove she attended the game, given that it is more likely than not that she did not pay?

Faced with a real person, one inevitably seeks much more direct information about the probability that he committed the misdemeanor in question. Courts regard "naked statistics" as too slender a basis for an opinion.

Perhaps there are justice-related reasons for this, and from a purely epistemic point of view, we should simply go with the statistics. Indeed I have argued exactly this in relation to the use of epidemiological evidence to prove causation on the balance of probabilities (Broadbent 2011b; Broadbent 2015b; Broadbent and Hwang 2016). But that argument applies only when we do not have access to better evidence, and are forced to make a decision anyway. It applies when we have good reason to treat the case before us like a ball that has been randomly selected from an urn. It does not apply in situations where this is a poor analogy. The question, then, is whether, when contemplating a given medical intervention, one is really in a situation that is relevantly analogous to drawing balls from an urn.

In my view, one cannot make a general claim about the confidence one ought to have in relation to a given intervention. This is not just because, as I indicated in the last section, evidence to support such a general stance is lacking. It is also because in particular cases, one might have evidence that defeats this general claim, even if that general claim were accepted (contrary to my argument in the previous section).

One dramatic way to demonstrate this is to imagine that one conducts a trial oneself. Yes, financial bias operates across the industry; but one should have quite a firm handle on whether one is employed by a university or a pharmaceutical company, and on who is funding the study. Is one seriously supposed to approach one's own findings with a prior probability that takes into account the influence of financial bias?

Some biases operate unconsciously, but again one might have good reason in some particular case to think that the effects of such bias are minimal or absent. Publication bias probably does not operate to exaggerate the evidence for effectiveness of penicillin, for example, because it would be much more interesting to publish a study showing that in fact it did not work.

In addition, one might have evidence that does not come from studies at all, but from experience as a clinician or patient. Stegenga argues, for example, that SSRIs are ineffective or barely effective, and points out that they are nonetheless widely used. However, many individual patients and clinicians feel that they have direct personal experience of effectiveness. Anecdote may be a poor kind of evidence, but what about one's own experience? If Stegenga himself took SSRIs and felt they had an effect, would he retain his view that they were ineffective? More to the point, *should* he retain it?

In my view it is obvious that he should not, but one could argue the point either way. On one way of looking at it, it is possible to be mistaken about effectiveness in one's own treatment; but on another, it is hard to see what more one wants of an anti-depressant than that it makes one feel that one is no longer depressed. The main point, though, is that it is very unclear how to apply a general fact about unreliable research in a case like this. It would be a bit like driving a car from Johannesburg to Cape Town and then wondering whether one really has arrived without a breakdown, given that it is an old car of unreliable make. Only a bit like that, because we can all agree whether you arrived in Cape Town or not, whereas individual patient reports are not accorded high evidential status in contemporary medical research. This is more of a problem for medical research than for the patients who regard themselves as having been made better. Hippocratic medicine prized individual reports of feelings wellness or unwellness, and as Stegenga argues, it may be that contemporary biomedicine could benefit from recapturing some of the Hippocratic way of doing things.

6.7 SHOULD WE BE MEDICAL NIHILISTS?

Wootton is Nihilist about past medicine, and Stegenga about contemporary Mainstream Medicine. Between them, they cover a lot of ground, since the methods of EBM that Wootton puts his faith in are the methods of contemporary biomedical research that are criticized by Stegenga. I have argued that neither Wootton's not Stegenga's Nihilism is compelling. We should not be Medical Nihilists.

In the case of Wootton, it is reasonable to agree with him about the ineffectiveness of past interventions, but not reasonable to be Nihilist on this basis. In the case of Stegenga, it reasonable to agree about the ineffectiveness of contemporary interventions, provided that one takes note of the circumscription of his thesis to the pharmaceuticals being developed by contemporary industry. His Nihilism applies only where the methods that are critiqued are the basis of the use of those interventions. I have also suggested that, even where this criterion is met, evidence about the particular intervention at hand may well render it ridiculous to factor in a base rate, as if one were randomly drawing a ball from an urn. It may simply make no sense to regard the intervention as a randomly selected sample of the product of contemporary biomedicine.

One of the problems I pressed against Wootton was that he neglects the possibility that medicine might successfully engage in inquiry even while failing to cure, and even while not achieving truth in inquiry but simply eliminating a series of falsehoods. The same point applies to Stegenga's Contemporary Nihilism. Stegenga writes:

> a central goal of medicine is to cure diseases or at least to pro-
> vide care for the diseased, and the most prominent class of
> interventions used to achieve this end today is pharmaceuticals.
> This is the subject of medical nihilism.
>
> (Stegenga 2018, 16)

I agree that this is a central goal; in fact, I think it is *the* goal. But I do not think that failure to reach this goal warrants despair about medicine. As long as we are meaningfully engaged in an inquiry with this goal as a purpose, we need not despair. We should not be Medical Nihilists.

Stegenga also indicates that he means Medical Nihilism to be synonymous with Therapeutic Nihilism. But the failure to reach the goal of cure does not even warrant Therapeutic Nihilism. As long as we are making progress with the project of understanding disease, for the purpose of curing it, we need not despair.

What is certain is that the progress with this project has been much slower than anyone would have wished, and remains slower than we want. This, however, is not a reason to think that medicine is worthless.

6.8 CONCLUSION

The view of medicine that I am developing owes a great deal to Stegenga's Medical Nihilism. It is for this reason that I have done the latter the philosophical honor of attempting to refute it. I want to close this chapter by exploring the commonalities and differences between Stegenga's Nihilism and my Inquiry Thesis.

The initial provocation for both Nihilism and Inquirism (forgive me) is ineffective medicine. Stegenga and I both feel the need to respond to the fact that *so much* of medicine either does not cure, or does not do anything beneficial at all, or harms. Stegenga's response occupies a conceptual framework in which the point of medicine is to cure, in a fairly strict sense of that term. Consequently, he concludes that medicine is, broadly speaking, a failure, notwithstanding occasional successes, and not denying their importance.

My fundamental difficulty with this reaction is that it does not make sense of the whole of medicine. Consider medical research beyond the assessment of clinical interventions. The detailed critique of research methods does not readily transport to observational epidemiology, which is not typically industry funded, and which usually seeks to identify harms. Yet we see just as much unreliability in observational epidemiology, perhaps more, as in clinical epidemiology (Vandenbroucke 2008). Science is difficult; financial interests exacerbate but do not create the difficulty. Or consider public health; as I have argued, the perspective of public health often directs that minimally effective interventions be adopted across the population—indeed, the less effective, the more widely they must be adopted to achieve a significant effect at population level. Consider surgery; financial interests operate here, but in quite different ways, and research methods are quite different from those upon which Stegenga bases his Nihilism. Consider, most importantly, the clinical encounter. What do people want out of this? What do they get? How are the two related? Stegenga and I agree, I think, that what they want is cure, and that what they get is often not cure. We disagree, I think, about how these are related.

The difficulty I have with Nihilism not making sense of the whole of medicine is removed if medicine is more than the purely pragmatic attempt to cure. If that is what it is, then Historical Medical Nihilism is warranted, and Contemporary Nihilism is warranted for all except Mainstream Medicine, and possibly for chunks of the latter as well—maybe even for all of it, since the pragmatic stance gives Stegenga a better basis on which to resist my suggestion that he sets the standards for effective intervention too high.

But if medicine is an inquiry, then it can fail to produce cure without failing altogether, and it can make progress even without making progress in curing. The *goal* of medicine, if it has one, may be to cure; but its *nature* may nonetheless be that of an inquiry. Standards for effectiveness are rather low, not because we don't want better cures, but because we do not expect them. Medicine does not *need* to cure us in order to be recognized as medicine. To do that, it needs to be engaged in an investigation into the nature and causes of health and disease.

At this stage, the alternatives are as follows. One can endorse Contemporary Medical Nihilism, as does Stegenga. One can endorse Historical Nihilism yet deny Contemporary Nihilism, as does Wootton (one can leave out his Whiggishness). Or one can look for a third way out. For me, there needs to be some explanation (better than the Whiggish one) of the persistence of ineffective medicine throughout history.

But maybe not for you. Maybe you are not impressed by the existence of ineffective medicine; maybe The Bullshit Objection convinces you, for example. In the next two chapters we will move away from Mainstream Medicine and look at other contemporary traditions. The fact that these persist provides, I believe, a further set of reasons for believing that medicine is not merely pragmatic, and that it is fundamentally an inquiry rather than a tool.

Chapter 7

Medical Cosmopolitanism

7.1 LOOKING ELSEWHERE

In his book *Cosmopolitanism*, Kwame Anthony Appiah describes an attitude that, he argues, is justified both in its own right and in relation to the consequences it would have if more people adopted that attitude. The focus of the book is ethics ("in a world of strangers," as per its subtitle), but the attitude includes epistemic, metaphysical, and practical stances as well as a moral one. While the focus of that book is our knowledge of and confidence in ethical judgments, inevitably it also concerns our attitudes to the beliefs of others, where they differ from our beliefs. This is inevitable because moral judgements often depend upon matters of fact.

I have no doubt that Appiah would advocate Cosmopolitanism about medical matters, since that view is intended as a stance we should adopt toward everything. I want to develop the application to medicine, which, I shall argue, contrasts with the attitudes more or less explicit in EBM and Medical Nihilism. It also contrasts with Medical Relativism, a view that is not espoused in these terms but which is sometimes evident in discussions of Alternative Medicine and Traditional Medicine, which are the subjects of the final two chapters of the book. I want to consider what attitudes are appropriate to non-Mainstream traditions, and I want to give some prior consideration to how to handle differences of opinion about matters that go to the heart of personal and cultural beliefs, before confronting those differences.

Cosmopolitanism was developed in the domain of normative ethics. Before we can apply it usefully to medicine, I need to develop it in various ways, and draw certain important distinctions. This will enable me to formulate Medical Cosmopolitanism clearly. Without this groundwork, that position will remain suggestive and imprecise. This means that the first part of this chapter is primarily concerned with Cosmopolitanism in general, seeking to extend and refine Appiah's account. Relatively little is said about medicine before section 7.6, when Medical Cosmopolitanism is formulated and defended against some initial objections. The remainder of the book puts the position to use, generating further support for it in the process.

7.2 APPIAH'S COSMOPOLITANISM

Appiah does not offer an explicit definition of Cosmopolitanism— one that says "Cosmopolitanism is the view that . . ." or "A stance/ person/attitude is Cosmopolitan if and only if . . ."

Looked at one way, Appiah's book amounts to an extended *implicit definition* of Cosmopolitanism: a definition by showing how the term is to be used. Perhaps that is wiser on Appiah's part than offering an explicit definition, given the difficulty of devising such definitions, depressingly coupled with their questionable utility. However, for my purposes, I do need to extract a set of explicit commitments. These might not exhaust the meaning of the term as Appiah uses it (I have commented already that attempts to provide necessary and sufficient conditions for the application of even the simplest concept are, in my experience, universal failures, with the exception of concepts in formal systems, which is one of the main reasons why formal systems are devised and used). Still, I need to draw out some elements of the position, in order to contrast them with the attitudes of (or implicit in) EBM and Nihilism.

In my reading, I discern at least four distinct stances within the position of Cosmopolitanism: a metaphysical stance, which is realist; an epistemic stance, which is humble; a moral stance, which is egalitarian; and a practical stance, encapsulated in the thesis of the Primacy of Practice. In sections 7.3 through 7.6 I make these stances explicit, before considering their application to medicine in section 7.7.

7.3 METAPHYSICAL STANCE

The *metaphysical stance* of Cosmopolitanism is broadly realistic: there is a fact about many of the matters about which we disagree, including some matters that are deeply embedded in cultural belief systems (whether there is a God, whether Jesus rose from the dead, whether ancestors play a role in causing or healing sickness). Where we disagree, it is commonly the case that at least one of us wrong, and maybe both. This stance is not a strong commitment to any particular metaphysical thesis, but rather a generally realist orientation, and it might be motivated by considering that it does the least violence to most of our actual beliefs, which typically seem to us to be beliefs about independent facts; they do not seem to be just figments of imagination. Of course, there are exceptions, such as matters of taste (which we will discuss later in this section). But where we really disagree, it is usually because we think there is an independent fact of the matter, and further that the other party is wrong about that fact while we are right. Cosmopolitanism adopts a defeasible presumption in favor of this minimal point of agreement between disagreeing parties, whose defensibility I now want to explore.

I mentioned at the outset the fact that ethical judgments inevitably have factual components is one reason that philosophers in general have recently pulled back from the old distinction between fact and value. Personally I think the fact/value distinction is useful, and ought not to be dropped too readily. However, what *is* clear is that even if a fact/value distinction can be maintained, almost every judgment we make concerns both fact and value. The attempt to distinguish wholly evaluative judgments from wholly cognitive ones is doomed, even if fact and value themselves are distinct.

The reason for this is that rationality itself is a normative matter. It is a matter of "oughts" and "shoulds." The assertion that two and two make five is *wrong*: not morally wrong, but wrong nonetheless, albeit in a different sense. It is wrong in a different way from the wrongness of doing something evil, but it is still wrong, and it is no coincidence of the English language that right and wrong, correct and incorrect, and so forth have both moral and cognitive applications. They indicate disobedience, breaking a rule—whether a moral rule, or a rule of rational thought.

Further, they both require that an agent makes a conscious choice. A fridge, on the other hand, cannot make a mistake; it can break down, but that is not the same thing, since the fridge does not face a choice about whether to function or not, and then make the wrong choice. Fridges cannot be either moral or rational, since they do not have choices. Even a computer cannot, at the fundamental level, make a mistake. It can run code that is flawed, or experience a physical malfunction, meaning that something physical happens that interferes with the function we wanted the computer to perform. But so long as a computer is, at bottom, just a load of switches, it simply runs; it cannot do this right or wrong, in either the moral or rational sense. Even an animal cannot be either rational or moral unless it has choices—and whether high mammals have such choices is of course a matter of debate. However, if an animal acts solely on instinct, or on the dictates of its nature, it is not acting rationally or morally.

A wolf participating in a pack hunt may simply be acting out its nature, and so long as it has no choice about what it does, it is not subject to evaluation for either the morality of its actions or their rationality. This is why we do not regard the wolf as morally wrong for subjecting its prey to terror, exhaustion, and a painful death. If, however, in the course of a hunt, a wolf has a choice about which way to run in order to maximize its pack's chance of success, then its choice may be rationally assessed, provided it is a choice that is conscious, in a suitably wolfish sense. Anyone with a domestic dog will find it plausible that wolves may sometimes have such choices, and thus be capable of rationality—of being right or wrong in this regard. Whether wolves have choices giving rise to morality is a more complex matter, and one that takes us off course. My goal is simply to emphasize the familiar point that there are both moral and rational "norms," and, while they are not the same, they have important elements in common, both giving rise to "oughts" and "shoulds," and both requiring that the agent makes a conscious choice.

This parallelism between reasoning and moral judgment can raise fears of, or alternatively lead you to endorse, epistemic relativism. This is the view that, for knowledge in general, the correctness of claims to knowledge is relative to something, and not determined by the fact of the matter that the knowledge putatively concerns. The line of thought goes as follows. If reasoning is governed by rules, and these rules have the same status as the rules of morality, then our knowledge of them is highly problematic. We have no very clear understanding of where moral norms come from, nor of why we believe them. Fundamental moral disagreements cannot be settled by appeal to empirical evidence; they cannot be settled at all in a decisive way, a way that shows beyond doubt that one party or both are mistaken. They require diplomacy, perhaps psychotherapy, and even then the outcome is not always positive. From inside, our own morality may appear to reflect matters of fact; but from the outside, the morality of others looks like mere opinion, and it is hard to see what others can do to change this. If we accept that the norms of reason are like the norms of morality, it seems that we end up abandoning the idea that it is possible to convince each other of matters of fact, even in ideal circumstances. If three and three make seven for you and six for me, then we cannot even agree whether we have any beers left from our six-pack after we each had three. We can check, but you may simply accuse me of stealing the last one. There are always ways to alter auxiliary beliefs so as to preserve the truth of a cherished hypothesis (Duhem 1914; Quine 1953). At this stage of growing inebriation, it may come to seem that not only morality, but everything, becomes a matter of opinion.

Why should it be a problem if we find ourselves being led to relativism? Relativism is plausible about *some* knowledge claims. For example, I might (perhaps rather oddly) claim to know that custard is delicious, but if you disagree, I will not (seriously) accuse you of an error. You simply have different taste. Relativism in matters of taste, at least within bounds, is commonplace. This is because we typically think of "tastes good" as being relative to the taster. If it does indeed make sense to say that I *know* that peanut butter and jelly sandwiches taste good, the fact I know depends upon me, and is a property of peanut butter and jelly sandwiches only in relation to me, and not of the sandwiches on their own without any mention of me.

In the literature, cases of this kind are often called cases of *faultless disagreement*. Although we disagree, neither of us is in any way at fault. We are not disobeying any independent "rules of taste." Rather, we are reporting facts about the way we experience custard, sandwiches, and so forth.

Relativism about the whole of knowledge is a serious philosophical proposition, and a difficult one either to defend or to refute. In its strongest form it implies that there is *no faultless disagreement at all*, and when applied to a particular domain, disagreement, or claim, it implies that there is no fact of the matter about that domain, disagreement, or claim, and thus that disagreement about it need not entail that at least one party is wrong. Although I am calling it epistemic relativism, this is both an epistemological and a metaphysical position. It says that our knowledge claims are neither justified nor made true by independent facts of the matters they concern. I am dealing with it in relation to the metaphysical stance of Cosmopolitanism, which asserts that there is a fact of the matter in at least a significant number of cases of disagreement. This is contrary to epistemic relativism, which, applied to disagreements, says there are no facts of the matter, and thus that it is not necessarily the case that at least one party is wrong: it is possible that they could both be right.

It is easy to see why such a position will be resisted. In the moral context, it is bad enough, since it gives us no basis for thinking that unspeakable atrocities are in fact unspeakable. But we are accustomed to that feature of life; we know that, when it comes to moral matters, there is no simple cure for those who are on a completely different track. It is more disturbing if it applies to knowledge in general (and not just of putative moral facts).

Relativism about knowledge in general threatens our grip on reality, or at least our sense of having a grip. I believe that custard is tasty, but also that it is a yellow viscous liquid and that among its ingredients are eggs, which come from chickens, which are domesticated birds that have co-evolved with humans over some few thousand years, and which ultimately share a common ancestor with us, which lived several million years ago, on the same planet we now inhabit, which is spherical and orbits the sun. And so forth. If you believe that chickens were put on Earth, which is flat, in their current form by a benevolent God concerned to provide us with eggs, then you disagree with me, and I with you.

It may be that the disagreement is faultless in the common moral sense, in that one of us has been the innocent victim of a sub-par education or partial information, or is mentally less capable than the other, or something like that. However, these are all faults, not in the moral sense, but in the sense of epistemic shortcomings. A *better* education, *fuller* information, *more* brainpower could fix these things. If the disagreement is epistemically faultless, then there is really nothing wrong at all, just as in the disagreement about whether custard is nice. You think it isn't, I think it is, and that's fine. No need to try to educate one another.

While we find that easy to accept for "Custard is nice," we find it much harder to accept for claims about matters we take *not* to be relative to any particular person's point of view. We find it particularly hard to accept for things where we believe that it ought to be possible to convince each other: whether there is a cat on the mat, for example. To accept that you and I can disagree about that despite good lighting, proximity to the mat, agreement about what a cat is, and so forth is to lose our grip on reality—or, more likely, to conclude that our interlocutor is losing their grip on it.

Epistemic relativism is difficult to refute, despite being open to some obvious objections. One of the best-known of these arises from asking: does relativism entail that there are *no* standards on knowledge claims—that anything goes? But that seems wrong: clearly not anything goes; we do not get airplanes to fly merely by believing that they will. There is some end stop to our mental lives, where we come into contact and sometimes conflict with the world, or with something, that does not necessarily do what we believe it will, no matter how hard we believe. Faith does not move mountains by itself. A relativist, however, might simply disagree, and assert that faith alone (unassisted by any Almighty) *does* move mountains. My assertion to the contrary is true for me, but the relativist's is true for her. More sophisticated relativists will point to the fact that merely saying that there must be a truth does not make it the case that there is one, let alone that we know about it (Bloor 2007; Bloor 2008).

This raises a second well-known difficulty. If all knowledge claims are relative to a point of view, and there is no faultless disagreement at all, then why is the relativist even disagreeing with me about my opposition to relativism? Should she not accept that, for me, relativism is false? But if relativism is false for me then it is false for everyone, otherwise it would not be false for me, since what it means for relativism to be false is that the truth of an assertion is not relative to the assertor. Paradoxes of this kind, leading to self-defeat for relativism, were discussed by Plato, and have put many people off relativism since. However, this apparent paradox does not settle the matter. The problem may be compelling for me, but it need not be compelling for the relativist, who might simply assert that this is a problem in my eyes but not in hers. If even the rules of reasoning are up for grabs, then there is no way to refute any position through reasoning, since reasoning is according to a set of rules that are agreed between the interlocutors, and in this situation there is no such set of rules. Contemporary defenders of relativism give more sophisticated rebuttals (Kusch 2006), and the question is not settled.

At this stage we may as well punch our interlocutor on the nose and remark that, from our point of view, that is a reasonable way to conclude the discussion. This is a common development in cases akin to that of the six-pack, discussed earlier: when people are drunk they often fail to see reason, and to agree on what seeing reason means. The result is not tolerance, but violence.

The development into violent conflict is definitely not inevitable. It is not prescribed by relativism. But nor is it proscribed. This is ironic, because the implicit motivation of much relativistic thought is to avoid the violence attendant on excessive dogmatism—the horror of ideology enforced, whether Nazi, communist, Christian, Muslim, or any other.

It is equally ironic, of course, if the anti-relativist uses the relativist's refusal to "see reason" as an excuse for the use of force. For the reason that anti-relativists, on their part, want to insist on the existence of immutable and objective rules of reason is that they see anarchy as the alternative. The relativist fleeing the violence attendant upon dogmatism ends up being unable to proscribe violence, and the anti-relativist seeking to prevent anarchy ends up prescribing it. Both defeat their original objections. Reason provides no way to enforce its own rules against those who do not accept them; if they are to be enforced, they require physical support.

There are many contemporary intellectuals who react strongly against relativism because it raises the specter of irresoluble conversations of the kind pursued in the previous few paragraphs, and worse, the prospect that anything goes, and that we may as well abandon intellectual life altogether, because any proposed rules of reason or acceptable behavior are open to immediate rejection. The concern is understandable, but it ignores the important point that anti-relativism likewise raises this specter. *Any* disagreement raises the specter of violence, because if we cannot agree about how to think, we will struggle to get each other to agree about *what* to think using only intellectual means. *No* intellectual position, mode of discourse, framework, or otherwise can protect us from violence, since any such position, mode of discourse, framework, etc. can be questioned or rebelled against, and in that situation, intellectual enforcement slips its anchor. No demonstrated application of shared rules of thought is possible without some shared rules of thought. Physical enforcement is the only available kind.

Thus epistemic relativism is neither to be lightly dismissed, nor casually endorsed. Rather, it ought to be understood as a philosophical provocation, like philosophical skepticism. Studying it can teach us a great deal about knowledge, reality, and other deep philosophical matters; but actually adopting such a position is to miss the interest of it altogether, just as much as dismissing it out of hand. Like radical skepticism, what makes radical relativism interesting is not that it can be defended but that it cannot be lived. Endorsing relativism misses the point that it cannot be lived, just as dismissing it out of hand misses the point that it can be defended.

This brings me back to the concern that, if reason is subject to rules rather like the rules of morality, we are led to relativism. Having discoursed on relativism, I want to make two points about that fear. First, the fact that a certain set of commitments lead, by logical steps, to relativism is not necessarily a refutation of those commitments. It does not necessarily provide us with a reason to reject this set of commitments. This is because it is possible that many of our commitments do lead to relativism, just as it is quite possible that much of what we think and say about knowledge leads by logical steps to a thorough-going skepticism. This just shows how little we understand, of ourselves, the world, and our intellectual lives; it explains why there is such a thing as philosophy. It does not mean that we must immediately abandon the position that seems to point in this direction.

Second, accepting that reason is governed by rules whose status is akin to that of moral rules does not, in fact, lead to relativism by any series of logical steps. This is because, as we discussed in section 4.1, there is no reason to assume relativism about morality. Such a position is a substantive one, not a direct consequence of noticing that we cannot easily settle disagreements about morality. The appeal of moral relativism is apt to fade when we notice that the inability to settle fundamental disagreements decisively is a universal feature of our intellectual lives, not confined to moral disagreement, and indeed even more disturbing when it concerns matters of reason.

The great fear that accompanies relativism is the fear of violence: the fear that we might face each other and find no way to reach agreement at all, even in matters that seem to each of us to be constitutive of rationality, and thus that we will end up fighting. Cosmopolitanism hopes otherwise. It is committed to the idea that there is a fact of the matter, and this makes it worth trying to find that fact out (as Appiah points out, relativism does not encourage conversation [Appiah 2006, 31]). The way to avoid conflict is through adopting an appropriate epistemic, moral, and practical attitude, as we shall now see.

7.4 THE EPISTEMIC STANCE

The *epistemic stance* is one of humility. Humility is not subservience—the complete abandonment of one's beliefs in the face of disagreement from persons or evidence. It is not even uncertainty, although the Cosmopolitan is likely to be less than certain about many matters on which others profess certainty. It is, rather, the general willingness to reconsider one's belief in the face of disagreement, which also means taking the differing view seriously.

The *epistemic* component of Cosmopolitanism is concerned with a strategy for forming true beliefs, not with preserving the feelings of others by being gentle in instances of disagreement. Appiah focuses on testimonial evidence arising from the beliefs of others, but I think it is appropriate to see this as a particular instance of a more general attitude to evidence of all kinds, because the fact that another person believes something different is only *epistemically* relevant if it amounts to a kind of evidence that one needs to consider. And I think it is at least as plausible that one should be open to reconsidering one's opinions in the light of empirical or other evidence as in the light of testimony: indeed, such an attitude is central to some stances that are opposed to Cosmopolitanism, such as EBM (I shall argue).

Thus I think it is reasonable to extend the epistemic stance to include being open to revising one's belief in the light of evidence in general, including empirical evidence, as well as testimonial evidence reflecting the differing beliefs of others. This is particularly important because in reality much that is called empirical evidence is in fact *testimony*. It may be testimony of the empirical evidence others have gathered, but that is still testimony (indeed, one of the most common kinds). Although we commonly call the results of an empirical study "empirical evidence," pedantic accuracy would demand that we say they are *testimony of* empirical evidence, for all of us except those who conducted the study.

Depending on how strongly it is expressed, EBM may differ from Cosmopolitanism in its epistemic component. If the evidence hierarchy is interpreted lexically (so that any amount of higher-level evidence trumps any amount of lower-level evidence), then EBM will be flatly incompatible with the injunction to be willing to reconsider beliefs in the light of new evidence. On the lexical interpretation of the evidence hierarchy, when we encounter new evidence that is not as high in the hierarchy as what we already possess, then we should simply ignore it.

Interpreted in this way, EBM flirts with irrationality. A more moderate and plausible understanding of EBM simply gives the higher echelons more weight, rather than discounting the lower ones completely in the presence of the higher. Taken this way, EBM is much more defensible. However, even on this more moderate interpretation EBM may remain incompatible with the epistemic stance of Cosmopolitanism. The Cosmopolitan openness to changing one's mind is all-encompassing. It will include the evidence hierarchy itself. To the extent that an EBM-er is unwilling to enter into conversation about the evidence hierarchy (with a willingness to change her mind about it, and not merely to evangelize), even moderate EBM remains incompatible with Cosmopolitanism.

Note that I stop short of declaring EBM entirely incompatible with Cosmopolitanism: it depends how the former is expressed.

An open-minded EBM-er might be open to reconsidering the evidence hierarchy at any time, but might continue to conclude, after each such reconsideration, that the hierarchy should stay. This approach would be compatible with Cosmopolitanism.

The EBM movement has indeed produced a number of different iterations of the hierarchy, and in this sense it has proved itself willing to reconsider its own hierarchy. The hierarchy appears to be akin to constitutional law: changeable, but not easily. However, this kind of changeability does not mean that EBM-ers are willing to reconsider their beliefs in every disagreement. The most common uses of EBM do not leave open the possibility of debating the hierarchy at every opportunity or request. Constitutions are not up for revision in every legal dispute, not even in principle. The evidence hierarchy is likewise not open for piecemeal revision, at least not in the hands of most practitioners. Moreover it seems antithetical to the spirit of EBM that the hierarchy should be revised in light of recalcitrant results; some might see this as ad-hoccery of the kind that EBM stands against. Individual EBM-ers might take a Cosmopolitan attitude, but equally they might not.

To summarize, if the evidence hierarchy is construed lexically it contradicts the Cosmopolitan epistemic stance, and if construed preferentially, it is compatible with it but also with its denial. Thus EBM is not committed to the Cosmopolitan epistemic stance. But committed EBM-ers could be Cosmopolitans, if they endorsed that stance.

One response to this discussion might be that the epistemic stance of Cosmopolitanism must itself be open to reconsideration, if it is to be consistently upheld. This is a reasonable retort, but not a problem for Cosmopolitanism. The Cosmopolitan must simply consider whether dropping this commitment is reasonable when it encounters disagreement. This discussion of EBM is a case in point.

7.5 THE MORAL STANCE

The *moral stance* is a view that, because we are all part of one humanity, and derive our moral worth from that humanity, individuals have equal moral worth. This view thus includes a metaphysical component, that we are all one humanity; it stands in contrast to, for example, the idea that racial or sexual differences are morally significant, or the idea that the beliefs one holds affect one's moral worth, even if their holders may be subject to moral judgment for holding them. The stance is broadly a kind of liberal egalitarianism. It is egalitarian because it derives moral worth from something that all humans share in equal measure, and thus assigns all humans equal moral worth. It is liberal because (although this does not follow from my compressed statement above) this worth is of a kind that prevents others, even in substantial numbers, from curtailing their ability to act as they see fit without a decent reason. The nature of this sort of decent reason is obviously a central topic for liberalism quite generally, and Cosmopolitanism does not have any particular contribution to make in this regard.

I do not have much to say about the moral stance here, because normative ethical questions are not the focus of this book. The main difficulty for the moral stance is that liberal egalitarianism might be thought to be a very culturally specific stance, specific to the contemporary West, and thus that thrusting it upon the whole of humanity is in itself rather un-Cosmopolitan. I do not venture either to press or refute this objection, because the moral stance is not strongly implicated in the themes of this book. However, I would be inclined to advocate a Cosmopolitan biomedical ethics as a natural companion to Medical Cosmopolitanism, and a Cosmopolitan biomedical ethics would certainly need to deal with this question.

7.6 THE PRACTICAL STANCE

Finally, the *practical stance* is encapsulated in Appiah's thesis of the *Primacy of Practice*. This states that agreement is often much easier to attain as regards *what* to do than *why* to do it, and further, that we should seek to reach agreement in matters of practical action first, pursuing agreement about principle, theory, and so forth only when practical matters are not pressing.

The effect of this thesis is twofold. First, if we are contemplating joint action and we agree about the Primacy of Practice, this reduces the number of other things that we have to agree on before we can act together. You and I don't have to agree about whether there is a God in order to agree that it would be wrong to embellish all the figures on the roof of the Sistine Chapel with walrus moustaches, even if one of us appeals to religion in explaining why this is wrong, while the other does not. To the extent that one prizes consensual action, this is an enormous attraction of adopting the practical stance of Cosmopolitanism.

Second, it recommends that, when faced with a disagreement about what to do, we do not immediately begin arguing about the reasons for our preferred courses of action, but rather start with some cases where we *do* agree, and see whether we can move from there. If the Primacy of Practice is correct, then the best way to reach agreement is typically not to seek agreement on underlying principles, since those are—according to the thesis—the hardest to agree on (besides their tendency to be vague, and to underdetermine action due to disagreement about factual matters). Rather, we should start with cases where we do agree, and seek to identify, in as specific a way as possible, and then by a process of comparison with the present case at hand try to deduce the values or facts that we disagree about. This may still not lead to agreement, but it improves the chances, and in doing so maximizes the areas where we *can* identify common ground—in contrast to the principled approach, which typically dramatizes differences.

Something akin to the Primacy of Practice is sometimes appealed to under other names in bioethics. For example, when the legalization of abortion was under discussion in the United Kingdom in the 1970s, several parties remarked that it would not help matters to first seek agreement on the exact timing of the beginning of a human life. The approach of the Warnock Report was to cut to the chase:

> Some people hold that if an embryo is human and alive, it follows that it should not be deprived of a chance for development, and therefore that it should not be used for research. . . . Although the questions of when life or personhood begin appear to be questions of fact susceptible of straightforward answers, we hold that the answers to such questions are in fact complex amalgams of factual and moral judgements. Instead of trying to answer these questions directly we have gone straight to the question of *how it is right to treat the human embryo.*
>
> (Warnock 1984, 60)

R. M. Hare generalized the problem (in discussing the same problem) as follows:

> If a normative or evaluative principle is framed in terms of a predicate which has fuzzy edges (as nearly all predicates in practice have), then we are not going to be able to use the principle to decide cases on the borderline without doing some more normation or evaluation.
>
> (Hare 1975, 204)

More recently, there has been interest in casuistry—reasoning from hard cases—instead of or as well as principle-based reasoning

(Forester 1996). Casuistry is how the English common law developed, and the common law is a serious and respectable intellectual achievement despite the fact that its principles originate (and are required to originate, when an authority is requested) in cases rather than its cases depending ultimately on principles. This casuistic character is being undermined in common-law jurisdictions by the recent explosion of statutory law (legislation is massively more common at the time of writing than it was even thirty years previously). However, in common-law jurisdictions, it remains important in many areas (for example, in property law). Moreover, as a mode of reasoning, is still important in common-law jurisdictions in statute-heavy areas of law, because statutes inevitably leave room for debate about how they apply to particular sets of facts that were not foreseen by the legislators, and the way that the law develops to fill these gaps remains similar to the way the common law develops.

To the scientifically trained mind, casuistry can appear to get things backward. How can it make sense to draw a rule from a case, or to reason that a given outcome is right in one case by appeal to its similarity to another? Where do these judgments get off the ground?

The same, or a very similar, question may also be asked of general principles, of course: where does a principle derive its warrant? Whether or not casuistry faces a special objection in relation to its starting point, it has one striking *practical* advantage, as follows. We are much worse at abstract thinking than at concrete thinking. Our grasp of principles often seems to be far more tenuous than our grasp of what it would be right or wrong to do in a given case. The intellectual task of deducing consequences from principles is one that we find far more difficult than that of identifying salient similarities and differences between cases. Thus both our grasp of the premise of a given chain of casuistic reasoning, and our ability to comprehend or execute it, seem far more robust than our corresponding grasp of a given general principle or our ability to reason from it.

To develop the example discussed previously, almost everyone agrees it is wrong to abort a fetus at full term while the mother is in labor. Almost everyone agrees it is perfectly okay to refuse to engage in sexual intercourse because one does not want a child (though not everyone). We could reason about intermediate cases by devising a principle, such as "any interference with God's plan is sinful," and considering them in light of that principle. Such approaches rarely yield intermediate positions, interestingly; in the Catholic case they notoriously proscribed even the use of condoms at the time that abortion was being debated in the West. Moreover, deductions are liable to be reconsidered, with previous deductions being denounced as erroneous. This reflects the fact that deduction is a difficult mental task.

Another approach to intermediate cases is to consider how similar they are to each of these clear ones, avoidance of sex and full-term abortion. Having decided one "hard case" this way, we can approach future hard cases with reference to our decision in the previous, not necessarily regarding that decision as immutable, but taking it as a guide, representing hard work done in the past. And so forth. For example, one may note that there is no conjunction of egg and sperm in the first case. From there, one may consider cases where there is fertilization, but no heartbeat. One may consider various differences and similarities between the full-term case. One still has the hard task of deciding which matter, but if one makes a decision in one case, one can build on it in the future, without shouldering the burden of deciding all future cases here and now with the utterance of a resounding universal principle. Principles may play a role in such reasoning, but the starting point of the process is not comparison of facts with principles, but comparison of cases with previously decided cases.

This is compatible with the claim that we are more prone to have intractable disagreements about principles, since it means that we can sometimes believe that we are committed to a principle that we are not really committed to, in the sense that we have not anticipated all its consequences or enumerated all its exceptions. One might, for example, commit to a principle that it is wrong to kill, without considering the killing of persons to save others, or to end pain, or at their command, or by way of self-defense, any of which might lead one to modify the principle. But asked whether a mother ought to be imprisoned who hits an intruder over the head with a sturdy bedside lamp as the intruder is about to set fire to a cradle containing a baby, we come to a firm and quick judgment that we are much less likely to revise through mere cogitation (although new facts, of course, could always cause revision). The fact that our judgments about cases are more confident and the fact that they are also more harmonious means that even when trying to reach agreement on principles, starting with practice typically yields a more direct path.

The Primacy of Practice thus has two things to recommend it: enabling consensual action even when significant areas of disagreement remain; and providing a faster route to agreement about principle, one that does not require us to offer a resolution of all future disagreements through consensus on a universal principle. These are both defeasible generalizations, of course: some cases might be so emotive that they cloud our judgment, or they might be symbolic, and so forth. There is obviously a place for principle in our cognitive lives in general and in our collective decision-making in particular. I do not mean to deny that. I simply mean that we should acknowledge how bad we are at abstract thinking compared to concrete, and that, as a rule, it is best to start with examples. Even philosophers find this rule useful.

7.7 MEDICAL COSMOPOLITANISM

We might break Cosmopolitanism down into a number of stances: an epistemic stance, a metaphysical stance, a moral stance, and a practical stance. I elaborate on each of these in turn in the sections below.

Cosmopolitanism is a general attitude, and Medical Cosmopolitanism is the application of this general attitude to medicine. The application is as follows.

- The metaphysical stance is one of realism regarding medical facts, such that where there are medical disagreements it is a reasonable presumption that at least one party is wrong (which does not imply that any other party is right).
- The epistemic stance is one of humility regarding medical knowledge, meaning a willingness to revise medical beliefs in light of disagreement with the views of others as well as empirical evidence.
- The moral stance is that we derive our moral worth from a single shared humanity, and that our moral worth is therefore equal, and further that this moral worth is such that it is wrong to curtail liberty without a good enough, undefeated reason.
- The practical stance is that practice is primary, meaning we are more likely to be able to agree on the correct course of medical action than on the reasons for it, and further that we should focus on reaching agreement in practical matters first when they arise.

Realism is deeply plausible about medical facts. If anything is a universal human experience, it is the experience of sickness: one's own sickness and the sickness of loved ones. I do not believe that my experience of holding a sick child in my arms differs in any significant way from the experience of a mother in rural Botswana holding her sick child in her arms. I do not have an argument to support this claim. However, it seems to me to be as secure a starting point as any other that I have relied upon in this book, perhaps more so. I do not say that the experiences are identical; only that they have a common element, and that this common element is significant, and characterizes that experience. Someone whose experience did not include that element would be having a different experience entirely.

Because of this shared component in the experience of sickness, whether of oneself or of others one cares about, it is not plausible to maintain that medical facts are entirely relative to culture, society, individual perspective, or similar. There are some medical facts that are universal for humans, as a minimum, and perhaps this universality is explained by their objectivity. For example, there is widespread agreement about which states count as healthy or sick, even though there is also considerable disagreement at the edges, or in the case of mental illness, or in specific cases, or in particular times and places where rather unusual traits have become highly valued or despised. We generally agree on health, and recognize it in ourselves and others reasonably well. Health is in part a medical notion. In respect of health, it is not plausible to be a thorough-going cultural relativist. My arguments in Chapter 4 support this contention.

If we accept that the goal of medicine is cure, as I argued in Chapter 2, then this provides another component of medicine that is not relative to culture, social norms, or individual opinion. Putting these together, it follows that there is general agreement about when medicine is successful at achieving the goal of cure, and about what it is that we seek from medicine for ourselves and our loved ones. We want medicine to return them to health, a state that we generally agree upon, from a state of sickness, which we likewise tend to generally agree upon.

If an evolutionary account of health is accepted, as advanced in Chapter 4, then none of this is very surprising. But the rejection of relativism about the fundamental goals and subject matter of medicine need not depend on that account. It depends simply on the fact, evident to anyone who has traveled even a little, and who has kept their eyes open, and who furthermore is not in the grip of an intellectual delusion, that the experiences of sickness and health, and the desire for relief from sickness and the return to health, are as universal as any human experiences can be.

The routes that medicine takes to reach its goal, and the explanations it offers for health and sickness, are highly culturally relative, of course. There is massive disagreement between cultures, times, and individuals in these respects. But none of these differences give rise to medical relativism. They are efforts to attain the same goal, and thus are measured against the same metric. If making sacrifices to the gods proves less effective at relieving HIV/AIDS than antiretrovirals, then antiretrovirals are to be preferred.

In the context in which I write, in the aftermath of discussions on decolonization of the academy in South Africa, it is sometimes asserted that we should seek "African solutions for African problems." Whatever the merits of this proposal, there is no sense in developing it in the arena of medicine. It would be immoral, in my view, to make an assertion of the nature of "African medicine for Africans and European medicine for Europeans." That is because sickness is not an African problem, but a universal one; and success in relieving sickness is likewise a universal matter. If something relieves sickness then it should be available to anyone, regardless of whether they are African, European, or of any other origin. These matters are revisited in Chapter 9.

Realism is often seen as allied with dogmatism, or even a species of it. I hope that it will be clear by now why I regard this as a misconception. The epistemic stance appropriate to medicine, and consistent with realism about it, is *humility*. Thus, to pursue the example, we should not simply assume that either sacrifice to the gods or antiretrovirals will be ineffective, and dismiss it out of hand. We should seek evidence, as the EBM-era insist. But unlike (most) EBMers, our conception of evidence must be likewise humble; it is no use being willing to test anything, but adamantine in one's refusal to consider whether one's evidential standards are correct or appropriate.

There is a tendency to see this as tantamount to relativism; this is the flip side of the misperception that realism is of a piece with dogmatism. It is not, and nor is humility of a piece with relativism. One can inquire with an open mind, and still come to a conclusion. One can likewise be open to revisiting the conclusion in light of new evidence, without being committed to revisiting it when there is no new evidence, merely because others do not agree and will not be persuaded. In my life as a dean, I sometimes have to enforce a decision that is not universally agreed upon, or even not accepted by a majority of staff. In doing so, I am not necessarily being dogmatic. I may have listened to all the various arguments, and then, when I am sure no new ones will emerge, and that a certain course of action is correct, I may make a decision. This is at least how I aspire to proceed. If new information, or a new consideration, comes to light, I may consistently review my decision. None of this implies that I am a dogmatist, or a tyrant, or that I do not listen, even if, in such situations, these criticisms are often leveled. One can be epistemically humble and still arrive at a decision, and act upon it.

In the medical case, a government can still legislate over the vaccination of children, or refuse to license certain practitioners as medical while endorsing others, or refuse to subsidize the prescription of certain treatments while subsidizing others, and so forth, while maintaining epistemic humility. What that stance requires is that they have approached the various treatments with an open mind. It further requires that they are hesitant to make strong pronouncements where uncertainty remains, as it often does in medicine. For example, there is no very effective cure for back pain, and many people claim relief from chiropractic, even if many others claim that there is no evidence to support its use, and some evidence for it being dangerous. In such a situation, epistemic humility probably prescribes a laissez-faire approach on the part of a legislature.

From the perspective of an individual medical practitioner, or of an individual patient forming an opinion, the approach is likewise a differentiated one. The doctor may strongly recommend that a patient take antiretrovirals rather than sacrifice to a god or pay a *sangoma* with the money that would be used to purchase the drugs. The doctor should be much less certain in advising a patient on whether to try chiropractic or back surgery. This may sound obvious, but the admission of uncertainty is far from common in medical contexts.

Because I am setting aside normative ethical matters in this book, I shall set aside the implications of the moral stance for medicine.

The final stance, the Primacy of Practice, gives rise to a continuum between clinical decision-making, through the decisions that a patient might make about treatment, to the big-picture assessment of the merits of a medical treatment, with a casuistic approach being endorsed for all.

The medical profession has embraced casuistic reasoning for as long as the legal profession, maybe even longer. The case study is a medical mainstay. The scientific comparison of two cases side by side to identify causal differences is a relatively recent arrival, and immensely valuable. But the single case study is still valuable. A detailed description allows a clinician to compare the case before him with that recorded, and consider the similarities and differences. This may allow him to draw inferences about how this case will progress if untreated, and how a given treatment may work. Perhaps they are not very strong inferences, compared to those warranted by an RCT. But the RCT is very narrow in its scope. A case study may be useful for many different patients, if they share characteristics that the clinician believes to be key with the case presented in the study. It may likewise be useful for guiding inferences about the course of cases that differ in different respects from the case in question. A case study is rich in detail, and any of these details may prove useful, even if the warrant they provide is not strong by the standards of an RCT. An RCT, on the other hand, provides a very strong inference in relation to a very specifically defined intervention on a very specifically defined kind of person—the more specifically defined, the stronger the inference.

Aside from the case study, clinicians naturally tend to reason casuistically in their own practices, developing a mental database of cases and comparing new cases to that database for similarities and differences. This is a mental task at which humans—and many other animals—excel, and as such it is a very good way for doctors to reason. It is a far better bet for daily decision-making than the conscious attempt to apply learned principles. It is the reason that experience is such an important part of medical training.

Clinical decision-making is thus commonly casuistic, and this is a good thing. The decisions that patients make are likewise casuistic: I had backache before and went to the doctor and they couldn't do much; this is a backache in a different place; I probably won't get much satisfaction at the doctor, then. I endorse such an approach in many commonplace cases. I likewise endorse this approach to the evaluation of treatments in a more general way. Rather than state some resounding principle, such as "We should only trust interventions that are evidence-based," or "We should only trust treatments that are based on sound science," I advocate considering treatments on a case-by-case basis, along with the claims that are made for them. In one sense this is not different from what EBM espouses in principle, although in another respect it differs markedly, since EBM (usually) advocates a universal standard of evidence, which amounts to a general principle.

7.8 CONCLUSION

I have unpacked Cosmopolitanism and sought to apply it to medicine, to yield Medical Cosmopolitanism. I have given some general reasons to accept this position. However, the proof of the pudding is really in the eating. I have criticized both dogmatism and relativism for their inability to provide nonviolent ways to resolve debates. No intellectual position can guarantee an escape from violence, but some may at least show us how to avoid violence. To establish the credibility of Medical Cosmopolitanism, then, I must apply it to areas of serious disagreement about medicine. In Chapters 8 and 9, I will do so, considering alternative medicine and the decolonization of medicine respectively.

Chapter 8

Alternatives and Medical Dissidence

8.1 MAINSTREAM, ALTERNATIVE, AND TRADITIONAL

So far, this book has focused on contemporary Mainstream Medicine, with some reference to its history, and more occasional reference to the existence of medical traditions outside the Mainstream. In this chapter and the next, I turn to those traditions.

At the outset I want to elaborate on the distinction suggested by the titles of these two chapters between *Alternative Medicine* and *Non-Mainstream Traditions of Medicine*, which, for short, I will call *Traditional Medicine*.

Mainstream Medicine is characterized by its position of global dominance, as a matter of social and political fact. Alternative Medicine exists alongside Mainstream Medicine, and is characterized by consciously and deliberately deviating from, augmenting, or rejecting some or all of the Mainstream framework.

Traditions of Alternative Medicine may be pitched as alternatives, whether for the whole or part of Mainstream Medicine. But they may also be pitched as *complementary*, implying that it does not seek to usurp Mainstream Medicine but merely to "complement" it. The implication is that there may be things that Mainstream Medicine somehow misses, or at least does not have the resources or time to address; or perhaps the idea is that there is an optional extra that one can have that is beneficial but not part of the standard package, like a sunroof on a car. Partly because of this variation in the extent to which traditions oppose themselves to the Mainstream, there is neither a standard nor an entirely satisfactory all-encompassing term for Alternative Medicine (Frank 2002). A common term is *complementary and alternative medicine* (Hansen and Kappel 2016).

I prefer the term "Alternative Medicine" because I think the core component of Alternative Medicine ignores the distinction between alternative and complementary. The core feature of such systems is that they *acknowledge the existence of Mainstream Medicine*, meaning that *they frame their offering as either compatible with or contradicting Mainstream Medicine*. The distinction between alternative and complementary is a matter of framing, and of the degree to which a particular tradition rejects aspects of Mainstream Medicine, as opposed to merely claiming to spot gaps.

At one end of the scale, physiotherapy occupies the border between Alternative and Mainstream. It is widely employed in Mainstream rehabilitation. However, its interventions are rarely evidence-based, and their theoretical basis is often not grounded in Mainstream theoretical frameworks. Many physiotherapists will use "dry needling," which has its origins in acupuncture. While there is a hypothetical mechanism for the effectiveness of dry needling that is compatible with Mainstream Medicine (and does not involve meridian lines) there is scant scientific evidence (of the laboratory kind) to secure these hypotheses, which remain largely speculative. Physiotherapists are also likely to be open to offering or, where they do not have the requisite training, recommending treatments from other therapies that are clearly alternative, such as chiropractic or homeopathy. Thus physiotherapy is not oppositional in its attitude toward Mainstream Medicine, and does not base itself on asserting any errors at the foundation of Mainstream; but it does at times go beyond what a Mainstream training would strictly justify.

At the other end of the scale, homeopathy outright rejects—in theory at least—fundamental elements of Mainstream Medicine. Homeopathy holds instead that "like cures like" and that the smaller the dose, the larger the effect. These two tenets are combined, so that homeopaths acknowledge dose–response relationships at higher doses, but maintain a special effect of extremely low doses when created by the proper dilution process. Thus two of homeopathy's (three) fundamental tenets arise from reversing a couple of tenets of standard pharmaceutical thought: that a substance producing an opposite symptom is a good remedy, and that the larger the dose, the larger the effect. The third tenet of homeopathy is that there is a vital force sustaining life and that healing involves stimulating this vital force. This notion is in tension with treatments that aim to eliminate the cause of disease with minimal assistance from the body—antibiotics, notably—but all the same, it is arguably less opposed to the fundamentals of Mainstream Medicine than the other two, since it has its roots in Hippocratic thinking, and is broadly consistent with the Mainstream notion of supportive treatment, even if expressed in a more picturesque fashion.

Thus Alternative traditions differ markedly in how they position themselves in relation to the Mainstream. What they share, I suggest, is that they all *do* position themselves in relation to the Mainstream. In this respect there is a strong parallel in the West between medical and religious dissidence (Porter 2002, 46), with varying degrees of opposition to the mainstream sect, but in all cases some sense that there is something missing from, or something to be gained through supplementing, the mainstream doctrine and practice. And just as there is a distinction between religious dissidence and wholesale religious difference, there is a distinction between Alternative and Traditional Medicine.

Traditional Medicine arises from a different and largely distinct conceptual and historical framework. At present most Traditional Medicine has come into contact with Mainstream Medicine and has been forced to acknowledge its existence, and in some cases modify itself as a consequence. But until relatively recently, there were medical traditions in the world that largely did not acknowledge Mainstream Medicine, in the same sense of "acknowledge" that I used above: they did not either agree or disagree with any tenets of Mainstream Medicine; they simply had their own theoretical and practical systems that had evolved in parallel. For example, Chinese medicine, Indian medicine, and Zulu medicine shared little history with Western medicine, the historical forebear of contemporary Mainstream Medicine. Thus contact with Mainstream Medicine tends to be a bit of a shock, on both sides.

The boundary between Alternative and Traditional Medicine is not sharp. Some Chinese medicine is practiced in the West as an Alternative. Those who consult it typically do so for the same reasons they might consult a homeopath: the view that Mainstream Medicine has somehow missed something. Acupuncture is usually classed in the West as an Alternative, despite being a characteristically Chinese intervention originally. However, acupuncture as practiced in the West *does* position itself in relation to Mainstream Medicine, for example by suggesting that Mainstream Medicine misses something that Chinese lore encapsulates, or else by arguing that there is a minimal but suggestive evidence base and that, in due course, the traditions will be shown to be compatible, and may merge. Both lines of argument have some plausibility, because acupuncture does yield persistent anecdotal reports of effectiveness in areas where Mainstream Medicine commonly disappoints.

Nonetheless, I want to retain the distinction, even if its boundaries are blurred. There are too many important differences between the situation of a Westerner who consults an acupuncturist in suburban London and the Zulu who consults a *sangoma* in sub-urban Johannesburg. Both may be well-heeled, but the reasons for the consultation are quite different. Both are likely to be dissatisfied with or suspicious of Mainstream Medicine, but for entirely different reasons. Crucially, the power relations between the cultures to which each belong are reversed: the Westerner is buying in Chinese medicine by choice, while the Zulu may feel that Mainstream Medicine has been thrust upon her and her culture. The situation of the dissident Protestant in a Protestant country is different from that of the Muslim in a country where Protestant Christianity is the state religion, even though Christianity and Islam are hardly historically isolated. These are good reasons not to lump the cases together, even if a principled distinction is hard to draw and easy to contest, as is usually the case for principled distinctions.

In this chapter I focus on Alternative Medicine. I want to identify the philosophical interest of the fact that there is such a thing, and to understand the reasons for and nature of *Medical Dissidence*. Some of the vitriol and mockery directed at Alternatives arises from a hope that they will be eliminated, and a kind of amazement that they persist. I will argue that Medical Dissidence is a nearly inevitable companion of Mainstream Medicine, at least in anything like its current form. The explanation I give arises from the epistemology of medical evidence, and especially the role of expert testimony, in Mainstream Medicine.

This picture makes the vitriol misplaced, and also misguided, since it is a poor strategy for achieving its end. A better strategy, I suggest, is provided by a Cosmopolitan attitude, one of tolerance and conversation, yet not of the kind of epistemic pluralism that degenerates into an unworkable relativism. This kind of attitude is not very fashionable at the moment, perhaps because it does not make for clear and strong positions, ready soundbites, tweets, and the other afflictions of a world that, despite record levels of both education and automation, seems to have less time for intellectual engagement than for almost anything else. Nonetheless, without the kind of conversation that Cosmopolitanism encourages, entailing a genuinely open attitude and a genuine willingness to be persuaded rather than an illicit agenda of conversion, the epistemological situation that gives rise to Alternative Medicine will continue. Alternative Medicine will persist until Cosmopolitans are in vogue.

8.2 FOUR BAD ARGUMENTS IN FAVOR OF EFFECTIVENESS

The big epistemological question about Alternative Medicine is whether it works: that is, whether it offers interventions with significant curative value. Given that I have argued that the core business of medicine is not cure, and that a tradition may be valuable despite being curatively ineffective, it will be obvious that I do not think this question is the only one to ask when assessing the value of a medical tradition. And given the epistemic stance of Cosmopolitanism, it will be obvious that I think that any assessment of what works must be open to discussion about the meaning of that phrase, "what works." However, the question "Does it work?" is commonly asked, and the sense of "work" is commonly determined not by conversation but by Mainstream Medicine. In this section I will consider four bad arguments in favor of effectiveness claims for Alternatives:

- that they would not persist if they did not work;
- that they fulfill other needs;
- that they are effective because they fulfill other needs; and
- that they are supported by scientific evidence or theory.

The least sophisticated argument for any given alternative therapy is: it persists, therefore it works. Thus Harald Walach writes:

> Homeopathy has some clinical effectiveness. If it did not, it would have died out.
>
> (Walach 2003, 7)

This conditional is hard to justify. Is there some general principle that all medical interventions that last for some period of time and do not die out, work? This seems highly doubtful, because of the number of medical interventions that persisted for centuries and that are now widely accepted not to have a positive effect: bloodletting, most notoriously. Of course, bloodletting does have *some* effect, making a patient weaker and quieter and so forth; and maybe this effect—which was not in fact beneficial—was thought wrongly to indicate healing. But this is hardly the line of defense that Walach is likely to take in relation to homeopathy.

A more promising line would be to deny the general principle but insist that, for homeopathy, the conditional is nonetheless true: if *homeopathy* lacked at least some clinical effectiveness, given the time, place, and all the other circumstances in which it was and is practiced, it would have died out. If this isn't clear, compare a different example. Following a car crash, I can assert that, had I not been wearing a seatbelt, I would have died; but I need not thereby commit to the claim that all seatbelt wearers survive, or anything of that sort. The conditional is true because of the circumstances of my crash and not because of a general truth about seatbelts.

However, even restricted to a particular practice like homeopathy, this conditional remains hard to justify, because of the general fact that a number of treatments, interventions, practices, and so forth do seem to persist despite the lack of a beneficial effect. Indeed, I have argued that much of medicine fits this tradition; I have argued that, in most times and places, most medicine has been ineffective. Yet it has not died out. Effectiveness is certainly not the only explanation for persistence: bullshit is another, and engagement in inquiry with the purpose of cure is a third. The inference from persistence to effectiveness is precarious, whether generalized or particular, unless there is some further consideration to show why this particular practice would not have persisted but for effectiveness.

A second argument is that the existence of alternatives shows that they fulfill *some* need that Mainstream Medicine does not fulfill. This may not necessarily be cure, but perhaps a psychological or emotional need—reassurance, being paid attention to, having someone listen to your story, or similar. This idea is made plausible in particular by the fact that attention to the individual patient is probably the single common thread most strikingly shared by Alternatives and most glaringly contrasting with the depersonalized nature of Mainstream Medicine. Porter writes:

> Once doctors became therapeutically far more potent, thanks to antibiotics and other magic bullets, they arguably abandoned the art of pleasing their patients. Armed with more effective weapons, they tended to forget the psychological significance and benefits of the close and trusting doctor/patient relationship patients expected.

(Porter 2002, 44)

This, combined with the antiauthoritarian mood of the age, may help explain the otherwise rather paradoxical resurgence of interest in Alternatives in the 1960s, when hopes were perhaps at their highest for continued medical progress, and had not yet begun to founder on the rocks of the continued intractability of viruses, cancers, and other modern medical disappointments.

The claim that Alternatives exist because they fulfill a need is plausible enough, but it is not a defense of the claim that they *work*, or are means to achieving any particular end. It would put Alternative Medicine on a par with music, which may fulfill a need in those who suffer, but is not thereby medicine. Fulfilling a need may not even be a positive thing, depending on the nature of the need. A smoker's emotional need for a cigarette might be better unfulfilled, all things considered. The picture of Alternatives as salves for the psyche speaks to the inclusion of a more caring component in Mainstream Medicine, more than to exploring Alternatives for their potential cures.

A third argument is an extension of the previous one: it is the idea that Alternative therapies *do* cure, but that they do so through mechanisms other than the ones envisaged by their theoretical frameworks. These could either be entirely accidental cures, or they could arise from the therapeutic benefits of fulfilling other needs as described in the second argument. On this view, Alternatives are wrong in theory but right in practice, with the practice then having the character of the old "specifics"—substances or procedures that have been discovered to have an effect but that are not integrated into a more general curative system. Of course, Alternative interventions *are* integrated into a system, but on this view, the system has no value; it is a fabric built up around a few or even a single chance encounter with cure.

For example, once you get through the voluminous theoretical structures surrounding them, the reality of osteopathy and chiropractic treatments may be thought to boil down to the same three or four "manipulations," resulting in loud "cracking" of joints in the back and neck. A lot of people find this satisfying and feel that it is beneficial for short- or long-term back or joint pain. Many people also try to "crack" their backs or necks themselves. Perhaps the rather small number of "cracking" procedures that are most commonly used by osteopaths and chiropractors are beneficial, even if not for the reasons envisaged by either, and even if Mainstream Medicine does not know the reasons either.

Even if homeopathy—to pursue another example—has no therapeutic effect through its remedies, the nature of a homeopathic consultation may have a therapeutic effect, perhaps *by* answering an emotional need that the patient has, to be listened to, taken seriously, and so forth. (This is subtly different from the psychic salve idea explored in the second argument, where an emotional need is answered but without therapeutic effect.) Maybe simply being listened to *is therapeutic*, for many of the ills of modern life; maybe no drug can cure the blues, but some kinds of human contact can, and Alternative therapies are one of the rare places to find this kind of contact in some contemporary societies. Or maybe being offered an all-encompassing explanatory framework that connects your warts to your insomnia to the fact that your teenage children are misbehaving at school is therapeutic; perhaps the idea that things make sense is helpful for us, even if it is really a mirage. This would be a "placebo effect," since the details of the treatment are not important at all, but rather the belief on the part of the patient.

One could also see the ritual of a homeopathic consultation as effective *regardless* of whether one believes that the remedies are effective. In that case, the consultation would be a little like a kind of hypnosis. Similarly, an atheist can acknowledge "feeling something" during a religious ceremony; the ceremony itself may be powerful, operating upon the participant's psyche in ways that are not understood, even if a participant does not believe. This would not be a placebo effect in the standard sense, since the details of the practice would matter; but they would still operate psychologically rather than through the mechanism they proposed.

In short, Alternatives might be effective in ways that they do not themselves understand.

This line would not be a defense of alternative therapies on their own terms. It is in effect an argument that they are either chance cures or very effective placebos, operating through the beliefs, expectations, and emotional reactions of the patient. In neither case are they acting through the mechanism that the theoretical framework of the therapy itself espouses.

This whole line of argument is not, however, a strong one. It confuses two things: an *explanation* for effectiveness that cannot otherwise be explained, and a *demonstration* of that effectiveness in the first place. It is really the latter that we want when we ask whether Alternatives work. If we can see that some Alternative works, but we do not see how this could possibly be explained in the way that its proponents urge, we might resort to this style of explanation, saying, "It obviously works, but we don't know how (and nor do the homeopaths/etc., despite what they say)." But this is a position of relative certainty, whereas the position we are actually in is one of uncertainty. We do not generally know that a given therapy is effective (and we may have reason to doubt it). In advance, on encountering a therapy, contemplating a treatment in exchange for hard-earned cash and precious time, or hearing a story from a friend or colleague about the wonders of this or that practitioner, we do not generally have a very strong reason to think that the therapy in question is curatively effective, unless we have some prior experience of our own (which I will consider in the next section). So this sort of argument does not give us much more reason to believe in effectiveness. It gives *some* reason—for example, we might think it is plausible that people who are listened to for an hour on a regular basis might do a bit better as a result. But, prima facie, it is not very plausible that this will have a significant effect on real and serious illnesses.

The most potentially convincing argument in favor of Alternative therapies is that they are supported by scientific evidence from clinical trials. However, this argument does not fulfill its potential to convince because for the most part Alternatives are not, in fact, supported by scientific evidence. This is part of what makes them Alternative. Alternatives reject, modify, or extend the theoretical basis of Mainstream Medicine, and so they are typically not supported by accepted biomedical knowledge, nor by reasoning from laboratory findings.

EBM is potentially Alternative Medicine's friend, then, because it de-emphasizes received wisdom and favors testing potentially effective interventions. However, for the most part, clinical evidence does not favor Alternatives any more than Mainstream Medical theory. In general, and with some exceptions (reminiscent of the "specifics" of Western medicine in the 19th century), Alternatives do not enjoy very much evidence from clinical trials; the positive evidence they do enjoy is typically of poor quality, from studies that are not compelling, either because of the venue of its publication (e.g., a study of mindfulness meditation published in a journal devoted to study of mindfulness meditation) or because the study design is either poor or simply not powerful (e.g., small, obvious potential confounders not controlled for) (Hansen and Kappel 2017). Sometimes, the therapies reject the relevance of such evidence; in homeopathy, it is the person and not the disease that is treated, meaning that trials of particular remedies or of effectiveness against particular conditions are ill conceived. However, such responses are flimsy. In this example, there is nothing preventing trials of the effectiveness of consultations for whatever was consulted for, as Hansen and Kappel point out (Hansen and Kappel 2016). It may be impossible to test the effectiveness of Belladonna 30C for a single ailment, since both the remedy and the dose are determined by the person as a whole and not just the ailment. But it would be possible to enlist a group of homeopaths in a double-blinded study and then randomly allocate placebo pills to half of them, and see whether the placebo group's patients perform markedly worse than the non-placebo group's. With patient consent it ought also to get ethical approval. No doubt other tests could also be contrived. Thus it is by no means inconceivable that a trial that does not violate homeopathic principles could be devised; it just requires ingenuity—and the will of the profession to submit to scientific evaluation.

Some commentators hold that the evidence "broadly favors" homeopathy (Vickers 2000, 49), and proponents of given Alternative therapies will be able to mention many positive results. However, the consensus is that the quality and value of most studies of Alternative therapies is low, and moreover for most therapies the evidence either is inconclusive or conclusively shows no effect over placebo. Singh and Ernst conduct a fairly skeptical survey from a thoroughly Mainstream standpoint and identify a few exceptions. For example, some herbal remedies are somewhat effective, which is compatible with the tenets of Mainstream Medicine; and some chiropractic and osteopathy is somewhat effective for back and joint problems, where the bar is low because there are so few effective measures for back pain (Singh and Ernst 2008). However, it is fair to say that scientific evidence, in general, and with some notable exceptions, is not a significant source of support for those who think that Alternative therapies, in general, are effective.

8.3 A COSMOPOLITAN APPROACH

Cosmopolitanism is a view about how to approach disagreements. The four-part recommendation is as follows. First, we should presume that there is a fact of the matter—a presumption in favor of the minimal point of agreement between disagreeing parties. Second, we should approach the disagreement with epistemic humility, which is a willingness to revise our own views in the light of evidence in general, including considerations advanced by our conversants. Third, we should regard our conversants as our moral equals. Fourth, we should usually put practice first, seeking to resolve disagreements about what to do before tackling disagreements about why to do it. We should prioritize practice over principle.

Applying this procedure to disagreement about Alternatives, the Cosmopolitan first recommends presuming that there are facts that are being disagreed about, whether they are facts about effectiveness or facts about the reasons for the effectiveness. This means presuming that at least one and possibly all disagreeing parties are wrong. Since all disagreeing parties tend to agree about this, I will not defend the presumption further here. In practice it will be easy to attain agreement on it.

Second, the Cosmopolitan recommends that disagreeing parties adopt the attitude that they are willing to change their minds about the effectiveness or theoretical basis of interventions, on the basis of considerations advanced by the other side. It is doubtful that such an attitude is common, to put it mildly. Those who attack Alternatives often use the language of charlatanism and quackery for practitioners, ignorance and irrationality for patients. Those who defend Alternatives adopt variable attitudes toward Mainstream Medicine, some equally aggressive. (On an antenatal course that I attended, run by a midwife, I recall that the opening image for the session on cesarean section was a cartoon of a surgeon wielding a chainsaw.) This means that I need to produce some compelling reasons for attitudes to soften. After all, it would not be very Cosmopolitan of me simply to insist that people change their attitudes because I say so. Thus in section 8.4 I will explain how the epistemic situation around medical facts leads to differing opinions almost inevitably. This mean that Medical Dissidence is probably here to stay.

Third, disputants about Alternatives should regard their interlocutors as having equal moral standing to themselves. This does not mean that no judgments may be advanced. It does, how-ever, reinforce the epistemic stance of humility, since it leads us to take our interlocutors seriously. Dismissing somebody's opinion merely because one learns that they believe in the effectiveness of ho-meopathy is not sufficiently respectful to satisfy the Cosmopolitan's moral stance. Dismissing somebody out of hand can be done, but the reasons need to be far more compelling—both the evidence against their view, and the reason for dismissing it rather than engaging or tolerating it. The moral stance is something of an idle wheel in this context because whether one is being unduly dismissive depends what one thinks of the rationality of the positions in play, and so I do not believe there is much headway to be made by promoting the moral stance of Cosmopolitanism in this context, except against the most obnoxious of interlocutors.

Fourth, Cosmopolitanism recommends pursuing agreement about practice over principle—that is, about what to do in particular cases first of all, and then working backward toward kinds of cases, and then to effectiveness of particular interventions or therapies, and only last of all to theoretical agreement. In section 8.5 I will argue that, in practice, this often happens. At least part of the controversy about Alternatives is manufactured, in the sense that it arises from principled opposition rather than concrete disagreements about what to do. Such disagreements do arise but their resolution is usu-ally a relatively obvious matter, and the probable effect of the contro-versy is actually to make sensible practical goals—the agreement of basic standards of practice, the detection of real charlatans—harder to attain.

8.4 EXPERIENCE AND TESTIMONY

At a basic level, we can draw a distinction between two kinds of evidence we might have for any treatment being effective: direct experience, and testimony. The distinction is not perfect, and nor is either source of evidence. In this section I argue, first, that direct experience of effectiveness is a source of evidence to the experiencer that is not easily overcome by testimony, and second, that anecdotal evidence of effectiveness is epistemically not so different from scientific evidence as is commonly claimed. The result is that it is difficult to convince people of the effectiveness of a medical intervention that you believe in, or of the ineffectiveness of one in which you don't, regardless of whether your belief is based on personal experience, anecdote, or scientific evidence. This makes Medical Dissidence (which I shall define) inevitable; and, I shall argue, it means that epistemic humility is the only reasonable attitude to adopt when engaging those with whom you disagree. It also suggests that a practical way to progress at least some disputes about effectiveness would be for disputants to try each other's preferred treatments.

Much of the evidence cited by those who favor Alternatives comes from the clinical experience of practitioners, and the corresponding testimony of patients. Many homeopaths believe strongly that homeopathy works (Frank 2002). Yet on its face, the testimony of practitioners and patients is a terrible reason to believe anything. Practitioners, in particular, stand to gain from perpetuating belief in effectiveness, regardless of its truth. Singh and Ernst suggest and imply a good deal of cynical charlatanism in homeopathy, in particular; they conclude their chapter by describing one particular homeopathic remedy as "the ultimate form of quackery," a pun on the fact the remedy comes from a duck.

But I suspect there is more belief in the process than this would suggest. The fact that it is chemically impossible to detect the difference between a homeopathic remedy and a sugar pill means that one could maximize profit margins by selling sugar pills, without—for example—extracting part of a duck and diluting it many millions of times. That process is expensive and if it could be clandestinely dispensed with, nobody would be able to tell from the end product. You can even have a quack with reference to homeopathy: somebody who sells sugar pills that have been produced the usual way. The notion of quackery is not properly referenced to what *works*, but to what ✗ is *believed*. Homeopathy is not quackery even if it is wrong; nor is it bullshit, even if it is ineffective, so long as its practitioners care about the truth of what they say, and believe that they are speaking the truth. Setting aside charlatans and dupes, there remain a substantial body of educated persons who practice or participate in Alternatives. These Alternatives dissent, in greater or lesser degree, from Mainstream medical science.

As I indicated at the start of this section, there are two reasons why this testimony cannot simply be dismissed as arising from the combination of reasoning post hoc ergo propter hoc and the phenomenon of self-limited disease. The first is that one needs to distinguish between one's own epistemic position, and that of one's interlocutor. If the interlocutor is telling you about an experience they have had, then, for you, that is testimonial evidence, but for them, it is evidence from their own experience. They are inevitably going to give this much higher credence than you, or they, would give anybody else's testimony. Perhaps they ought not to be certain, because their knowledge is not a direct experience but a causal inference from their experiences. Nonetheless, even causal inferences can feel like something. I am sure that I feel the effects of my morning cup of tea. I am reasonably confident of having felt effects from chiropractic manipulation, physiotherapeutic massage, and dry needling. My confidence is less than in the effectiveness of tea, but nonetheless it is high enough for me to take it seriously.

The second reason that personally experienced effectiveness—less politely, anecdotal evidence—cannot be dismissed is that it shares with scientific evidence the fact that it comes to the vast majority of its recipients through testimony. This means that, apart from one's own experiences, assessment of any evidence of effectiveness involves two separable assessments: one of the force of that evidence if the report of the evidence is true, and the other of the truthfulness of the report. There is no doubt that scientific evidence is much more compelling than anecdotal evidence of personal experience in cases where we are equally confident in the accuracy of the report. However, it is not reasonable to ignore the other element of the assessment, namely the assessment of the accuracy of the report. A failure of trust, even a relatively slight failure, may lead one person to reject evidence that another regards as rationally compelling.

It seems very plausible that failures of trust are a large part of the explanation for various kinds of contemporary skepticism about science, such as skepticism about human-caused global warming, or vaccine hesitancy (Goldenberg 2016). In some parts of sub-Saharan Africa, one source of resistance to the use of antiretrovirals against AIDS arises from suspicion that the drugs, or white doctors, are actually causing AIDS, as part of a white plot to rid the continent of black people.

It is much less clear that accusations of irrationality will stand once you take into account the possibility of different initial attitudes of trust. The Australian comedian Tim Minchin has a joke about Alternative Medicine. In one of his sketches, he says (roughly):

There's a name for alternative medicine that works. It's called medicine.

It's a funny joke, but it is not an insightful piece of epistemological analysis (nor indeed an accurate assumption about Mainstream

Medicine, which contains substantial areas of ineffectiveness). His implied claim that Mainstream Medicine works is based on the testimony of others, combined with his own personal experience. Minchin's personal experience is in the same boat as those who claim they have personally experienced effective Alternatives, so it does not differentiate Minchin's claim from the claim of those who defend an Alternative. The other evidence he relies on, and by far the largest portion, is testimony. It is testimony of scientific studies in some cases: journal articles, summaries of articles online, and so forth. However, in most cases, it is not even that: it is testimony from people who are experts in the field and say that scientific studies have shown effectiveness. One can read a paper in a top medical journal every day for a year and still absorb only a fraction of the evidence that is constantly being produced. Science in general, including bio-medical science, relies massively on trusting others to accurately report, analyze, summarize, and operationalize findings. In short, we rely on experts.

The difficulty with expert testimony, as Goldman points out, is that only other experts can readily assess their expertise. I might be able to tell that someone who professes to be capable of helping with my sore back is talking nonsense, but I might not. I am not an expert in backs.

What compounds the problem in the case of Alternatives is that academics, science journalists, and others who are in a position to articulate and promulgate their views tend to share a confident attitude toward the testimony that one finds in scientific journals, at scientific conferences, and so forth. They tend to feel that they are reasonably well equipped to understand what is said and to assess it, or, where they cannot assess it, they feel they know where to turn in order to get help with the assessment. Thus if I want to find out about the long-term effects of breastfeeding, I can email a senior epidemiologist whom I happen to know, a former president of the International

Epidemiology Association, who has spent considerable parts of his career researching this. If I want to know about asthma, I know someone equally senior that I can ask. I have no medical training, but if I wanted to, I suspect I could get extremely expert opinions on more or less any medical controversy of the day. Moreover I feel I would be able to assess these opinions in a critical way. I am in a special position, of course; not everybody spends time thinking and writing about epidemiology, healthcare, medicine, and so forth. However, those who *do* spend time thinking and writing about these things typically also end up with a network of connections.

The majority of people, however, have no such connections. They must find their way through a sea of clamoring voices. Some are obviously the voices of charlatans, but many are not. In this situation, it is not irrational to decide that those who clamor on behalf of science are wrong about this or that. After all, there are plenty of areas where the scientifically founded advice changes, meaning that at least some of what is currently recommended must also be wrong. Since admissions of fallibility are rare, despite this track record, suspicion is reasonable on this ground alone. Couple this with testimony from trusted persons of personal effectiveness of some Alternative, or indeed of mistreatment at the hands of Mainstream doctors, and the epistemic position of the Alternative grows.

Then suppose that one decides to pay the Alternative a visit, and is impressed: after twenty years of back pain, the chiropractor examines the X-rays, gives an analysis that accurately fits your experiences, and that is far more comprehensive than what the orthopedic surgeon ventured, and then moves your joints in ways that bring some pain relief along with movement that you have not felt in years. You notice that weakness in your arm or leg seems to have diminished. In short, the initial results are impressive, and you consider returning for some more treatments, even though they are costly.

Against this background, you pick up Ernst and Singh's book, and read the chapter on chiropractic. It warns of dangers whose extent is not known, it argues against effectiveness, it portrays a profession with huge overreach in what it claims it can do, and it relates an anecdote about someone dying at the hands of a chiropractor. None of this chimes with what you have experienced; your chiropractor seemed focused on your back, mentioned that she might not be able to assist with certain sorts of problems, and insisted on seeing scans before manipulating. You also have reports from trusted people that this person has really helped them. You have no evidence of injuries, and indeed it seems that nor do Ernst and Singh, since they gesture to unknown harms. Given your experiences, it is hardly irrational to conclude that you do not agree with the view of chiropractic advanced in that chapter, which is in a book purporting to set out the scientific facts in relation to a range of Alternatives. You do not thereby decide to reject science itself. You simply do not trust the research, impartiality, or intelligence of these authors.

It is certainly not a rational obligation to accept what someone says as gospel simply because *they* tell you it is scientific. Nor is it ever a rational obligation to ignore your own experience. And, of course, if you resort to Google, you will find a very mixed bag of evidence, all of it testimony by the very fact it appears on the internet, and thus subject to a reasonable level of scrutiny and some suspicion (even photographs are manipulable). You decide to go back to the chiropractor, uncertain whether or not there is a scientific basis for chiropractic, but impressed by your own experience and the testimony of those close to you.

Are you being irrational? I believe not. Nowhere in this story is there stupidity or irrationality, or even credulity. Unless you are yourself involved in the world of biomedical research, you will adopt a reasonably critical attitude to all evidence, and you *ought* to adopt that critical attitude even to evidence that is accompanied by assertions that it is scientific, just as you ought to be critical of assertions accompanied by claims like "this is the truth" and exhortations like "trust me." Perhaps one is even obliged to be *more* skeptical when one encounters this kind of content-less amplification. Indeed, against this background, the credulous stance appears to be Minchin's, since he seems to be under the impression that Mainstream Medicine is universally effective and stands on an unproblematic scientific base, when, as Stegenga shows and as we have discussed at length earlier in the book, the scientific base is often shaky, and effectiveness is definitely not universal.

I think that there is no principled epistemic difference between the person who rejects the effectiveness of all Alternatives, and one who accepts the effectiveness of some Alternatives. Both can be rational, and both can be irrational; either could ignore good evidence that they ought to consider, or make mistakes. It depends on one's personal experience, combined with the access one has to scientific evidence. This access remains extremely limited, despite the existence of the internet.

Given epistemic parity between believers and disbelievers in Alternatives, the only reasonable approach to disagreements is epistemic humility. This does not mean believing that one is wrong. It means being willing to consider whether one might be wrong. The other side will offer reasons, and you must try to understand them, not merely refute them. You must also offer reasons for your contrary beliefs, and the other side must try to understand you. In the end, it may come down to something irresoluble, such as a generalized sense of trust or mistrust in the medical profession. However, in many cases, it may come down to something far more practical: the pro-Alternative will have had their back sorted out by a chiropractor, or their eczema healed by a homeopath, or their headaches cured by an acupuncturist, whereas the anti-Alternative will either have had bad experiences or no experiences of Alternative therapies. In such cases, one course of action would be for the pro-Alternative to invite (perhaps even pay for?) the anti-Alternative to try the favored treatment, with the favored practitioner. Another would be for the anti-Alternative to walk the pro-Alternative through the body of evidence that convinces her not to believe. If both these things happened, we might still have disagreement; but we would probably have more understanding. And we would probably have fewer instances of name-calling.

8.5 THE PRIMACY OF PRACTICE

Alternative Medicine is growing (Eisenberg et al. 1998; Ernst 2000; Eardley et al. 2012). Porter writes:

> This astonishing resurgence of alternative and complementary medicine throughout the West, among young and old, rich and poor, people of all ethnic and religious background and all points on the political compass, shows that regular medicine has ceased to convince the public of its own creed: that it is the only, or the best, means to cure their ills.
>
> (Porter 1997, 688)

However, Porter describes a softening in the attitude of the Mainstream over time; in the early 1980s the British Medical Association was threatening disciplinary action against doctors cooperating with osteopaths, while by 1988 the Royal Society of Medicine was recommending "bridge-building" with alternative medicine (Porter 1997, 689–90). Though I have no proper evidence for this, I suspect that attitudes may actually have hardened again since Porter's book, which predates the EBM movement and the powerful public attacks on homeopathy and chiropractic in partic-ular that occurred during the 2000s (Singh and Ernst 2008).

Nonetheless, there is no doubt that Alternative Medicine remains controversial, and most likely the controversy arises from a perceived threat posed to Mainstream Medicine. However, in *practice*, the ex-istence of multiple traditions does not force either practitioners or patients to choose nearly so often as this controversy might suggest. The pressure to choose often comes from above: it often comes, not from conflicting ideas about what to do in a particular situation, but from proponents or defenders of one or another tradition. Theories can conflict while their practices do not, or to a larger extent than their practices. Nothing prevents practitioners from mixing and matching, apart from professional codes (which is probably one function of these codes), and nothing prevents patients from consulting multiple traditions. To some extent, the controversy is *manufactured*.

When I took my daughter for her latest round of vaccinations, the nurse suggested a homeopathic suppository in the event of a fever later, on the basis that the baby was under six months so too small for acetaminophen (paracetamol). Here was a Mainstream nurse, administering vaccinations, who also apparently believed in home-opathy. Or maybe it was a sop to appease worried parents, and she did not believe it would work; but even a sop like this would not be tolerated by those who really believe homeopathy is charlatanism. The suppository costs money, and the parents might be pulled in to homeopathy and end up dangerously eschewing proven medical treatments in favor of mere sugar pills, and so on and so forth.

What is more, the nurse did not know that two weeks earlier the child had been hospitalized with a chest infection, and had received plenty of acetaminophen during her stay, and moreover that we had been prescribed some to take home.

The whole thing made me chuckle. Here were two contradictory traditions being used together (homeopathy and vaccination are especially at odds), and moreover contradictory advice even within the Mainstream tradition (about acetaminophen use). What is more, I happen to know that the notion of a homeopathic suppository is virtually incoherent from the point of view of homeopathic theory. Thus the episode showed me first hand that medicine in practice is not monolithic, and that the oppositions that are sometimes set up (homeopathy vs. vaccination; acetaminophen use in infants vs. not) are at least to some extent constructed or imposed from above, and thus not a necessary feature of medical practice.

One might object at this stage that the nurse was simply being inconsistent. She ought not to have mixed and matched in this way. The two traditions she appealed to are logically incompatible, whether she recognized it or not.

But this is a confused objection. The traditions might be incompatible, but that does not mean that the two interventions she employed might not both work. A Western boxer's left hook and a roundhouse karate kick might both work even if the philosophies underlying them are not compatible. The nurse cannot coherently *believe* both traditions; and she cannot simultaneously believe the *explanations* for why each intervention might work. But she can believe they will both work; all this requires is that at least one of these explanations is mistaken.

Thus the practitioner is not logically compelled to choose between traditions; the nurse was not necessarily being feeble-minded. Yet some proponents, and more critics, of Alternative practices either assert or imply that a choice must be made. This suggests that to some extent the controversy about alternatives is manufactured by people who want to either promote or prevent them.

The Cosmopolitan is committed to the Primacy of Practice: the view that disagreements about what to do can be resolved without resolving disagreements about why to do it, and further, that disagreements about what to do are often easier to resolve than disagreements about principles; and thus that the best strategy, for reaching agreement, is usually to start by trying to find agreement about practice.

In the context of Alternative Medicine, the Primacy of Practice may prove particularly useful. Those who detest Alternatives are frustrated by their existence, and those who endorse them are sometimes highly critical of Mainstream Medicine, though by no means always. Focusing on practice resolves many of the points of disagreements. Homeopathic doctors in Germany are relatively common (Frank 2002)—medical doctors who are also homeo- pathically trained, and practice homeopathy alongside Mainstream Medicine. Seen one way, they are paragons of irrationality, even worse than pure homeopaths since they appear committed to two conflicting theoretical systems. But this is not the best explanation of what is going on. A better explanation is that they are adopting a pragmatic attitude, and using whatever seems to be most appro- priate in the circumstances.

In doing this, they are endorsing the Primacy of Practice, since they are proceeding to act without first resolving all theoretical problems. A real problem arises only if they feel pulled in two dif- ferent directions. This may happen. But for many cases, there will be no question in their minds about the correct kind of treatment, especially in acute disease. Where there is a clearly effective inter- vention, they will use it, or refer appropriately: hydration for cholera, antibiotics for tonsillitis, physiotherapy for shoulder rehabilita- tion after an injury. But for the very large number of ailments and complaints that are less susceptible to Mainstream treatment, or even to Mainstream explanation (e.g., various kinds of unexplained pain; insomnia; eczema), they may try homeopathy. This is nonsense from the perspective of either Mainstream or homeopathic theory. But both theories are at best partial; we know this by the mere ex- istence of unexplained symptoms. And from a curative perspective, both theories are likewise at best partially successful, since neither yields cures for every ill.

The Primacy of Practice provides a partial cover for different medical traditions to operate alongside each other without coming into conflict. Provided they do not disagree about what to do, they need not be enemies in practice, even if their theories are enemies. Such a pragmatic arrangement is quite common in the real world and goes a long way to explaining the persistence of Alternatives despite incompatibility with a powerful Mainstream. Although there are people on both sides—advocates of Mainstream Medicine and advocates of various Alternatives—who adopt much more aggressive postures, these are in part just postures.

The Primacy of Practice is one way to reduce the temperature of debate about Alternatives. However, it is not a universal prescription. It does not remove all disagreements, only reduces the number of those that we should feel compelled to resolve. The remainder, I suggest, need to be approached with an attitude of epistemic humility.

8.6 CONCLUSION

In Alternative Medicine, claims of efficacy are not well supported, but there is no obvious qualitative difference between the kind of evidence that supports them, and the evidence that supports many Mainstream treatments. Both depend to a very large extent on the judgment of an experienced clinician. From the point of view of everyone else—patient, onlooker—this means relying on the testimony of an expert. And it is notoriously difficult to choose between experts who disagree (Goldman 2007). The fundamental epistemological problem posed by Alternative Medicine is not how to detect the effectiveness of a treatment; we have a good handle on that. It is the problem posed by experts who provide testimony on topics that we do not understand, and who disagree with each other. Where these experts are clinicians relying on their personal experience in the clinic, it is extremely hard to see how to access, let alone assess, this evidence.

In a situation where Mainstream medical science is obviously incomplete, and most probably mistaken about at least some things, personal and clinical experience counts for something. If you believe that your longstanding backache was cured by two visits to the osteopath, you will be a convert. Osteopathy is one of the therapies for which there is some degree of Mainstream-acceptable evidence of effectiveness, but that is not why you will be a convert. You will be a convert because it fixed your back.

You will also be a convert if you visit a Reiki healer and achieve the same effect. It is hard to argue with someone who has seen something with their own eyes, and the healing experience can be like this. Vickers suggests that evidence for homeopathy must clear a higher standard because of the Humean point that, no matter how trustworthy the source of some testimony is, no testimony can outweigh the evidence of our senses. He takes chemical knowledge to fall into the latter category, since it is empirically based. But a more appropriate use of Hume's argument concerns the direct experience that many people at least *feel* that they have of the effectiveness of medical treatments, whether Mainstream or Alternative. My own knowledge of chemistry is entirely based on testimony, and even most chemists' knowledge of chemistry is testimony-based except where they have experimental expertise.

I have already mentioned how Hume himself might have responded. One cannot experience effectiveness, since one cannot experience causation. But this does not change the psychological conviction we feel when we encounter something that we think works. It also does not change the fact that, for the majority of us (even scientists), scientific evidence is in fact testimonial evidence, not direct experience. An experience of what *really* appears to be causation is always going to be hard to dispel with what someone says they observed in some big study that you don't really understand, and whose reporting you couldn't verify for accuracy even if you did. Against this background, epistemic humility is the only reasonable stance.

It is also the only way to root out the actual charlatans and dupes, who gain considerable cover by being lumped alongside practitioners who believe in what they do, and do not seek to exploit stupidity, gullibility, or irrationality in their patients. Calling all practitioners con artists and all patients suckers provides cover for the real con artists, and no help for the suckers.

Chapter 9

Decolonizing Medicine

9.1 MEDICINE AND POWER

Medicine exists in many forms, in many places and times.[1] There are different medical traditions existing in the world today. One tradition, which I have called Mainstream Medicine, is uniquely dominant. Even though that one tradition is, in fact, not at all monolithic, and includes many different and sometimes conflicting practices, it does nonetheless constitute a recognizable social institution. That institution has global reach, attracting levels of support and reliance from both states and individuals that is unequaled by any other tradition.

What does this mean for other traditions? In section 1.4 I briefly illustrated the interaction of Indian, Chinese, and African medicine with Mainstream Medicine, which is of primarily Western origin. There I emphasized that the predominance of Mainstream Medicine is in the first place the result of political dominance. This does not contradict the suggestion that Mainstream Medicine is uniquely effective, but it does militate against an easy inference from ubiquity to effectiveness.

---

1. Parts of this chapter reuse and develop ideas first published in *The Conversation Africa* here: https://theconversation.com/it-will-take-critical-thorough-scrutiny-to-truly-decolonise-knowledge-78477.

The relationship between power and knowledge is uncomfortable. On one side of the coin, knowledge yields power, to the extent that it yields the ability to do things that people without that knowledge cannot. However, an elementary philosophy of science course will (disabuse) anyone of the notion that a theory's truth can be established merely by pointing out that it yields practical power. The history of science is littered with theories that were successful, in the empirical sense, and appeared to yield power in the sense of guiding us to do things that help us achieve our goals, and yet have subsequently turned out to be false. Newtonian mechanics is possibly the most dramatic example. It is strictly false, but it yields practical power, allowing us to calculate the trajectories of missiles, for example.

This makes for uncomfortable reflection on our own science, including our medical science, and this discomfort increases when we look at the other side of the coin, which is that power directs knowledge in various ways. Marx held that those who own the means of material production likewise own the means of knowledge production. If you are in a position of power, you can force people to at least say that they agree with you. You can control what is taught in schools, and you can determine by force which arguments are deemed to be successful and which considerations compelling. I expect that most contemporary readers of this book will regard Mainstream Medicine as epistemically and practically superior to other traditions, even if it has flaws, and even if other traditions have a few valuable specifics or ideological contributions. Yet in the next century or so, perhaps even during what is left of my lifetime, it is quite plausible that China will rise to a position of global political dominance. In that case, Chinese medicine might likewise become more dominant, and its effectiveness might likewise be assumed by educated elites who have little medical or scientific education, but trust the knowledge-authorities of the day.

Thus it is quite possible that, in a hundred years, most readers of this book will regard Chinese medicine as epistemically and practically unique, with Western medicine being an interesting collection of techniques, some useful, some harmful, some simply eccentric; but all ultimately based on a view of health, disease, the human body, and the world that is fundamentally incorrect, albeit charming. Those readers will likewise acknowledge that political forces are partly the reason for the dominance of the medicine they espouse, but they will not doubt it for that; they will struggle to believe that politics could be solely accountable for its position of global dominance, and will argue that it would certainly not have retained this position in the face of constant testing in use and investigation in scientific institutions, if it were fundamentally flawed. This book may be regarded as containing flickers of prescience for not simply dismissing or ignoring Chinese medicine, but it will still be mired in the prejudices and circumstances of its day, and thus of limited usefulness.

These thoughts are uncomfortable for Mainstream Medicine. It is no part of medical teaching that we know the heart pumps blood because people who believed this were militarily successful in previous centuries and were able to disseminate their view and quell dissent. Mainstream Medicine does not explain its rejection of Zulu ancestral views of illness causation by reference to Zulu military defeats. Yet the historian and the anthropologist will point to such events as among the causes, and perhaps even the primary causes, of the content of medical knowledge in contemporary South Africa.

Arguments of this kind have been made in great and compelling detail by a number of writers. They are closely associated with Marxist thought. Karl Marx famously maintained that:

The ideas of the ruling class are in every epoch the ruling ideas.
(Marx, Engels, and Arthur 1970, 64)

Michel Foucault thought that power, truth, and knowledge are closely related, or even the same thing, which he sometimes called power/knowledge (Gutting 2014). A little later, those in the "Strong Programme of the Sociology of Scientific Knowledge" (see, e.g., Barnes and Bloor 1982; Bloor 1982; Bloor 1991; Bloor 2007; Bloor 2008) used the history of science to defend a sophisticated epistemic relativism on the basis, fundamentally, that the fact that people or groups hold particular beliefs can be explained by reference to social factors, without reference to the truth or falsity of the belief. Past scientific beliefs that are now considered to be false can hardly be defended on the grounds of their truth; the explanation for their popularity must lie in something else, and, they argue, the same goes as much for contemporary scientific beliefs.

In section 7.2 I explained my attitude to epistemic relativism, which I see as a philosophical challenge, and not a position either to be endorsed or rejected out of hand. In my view, relativism about the history of chemistry is one thing, relativism about the effectiveness of antiretrovirals another. It is one thing to profess epistemic relativism in a seminar room, another to profess it in a hospital; that is the difference between defending a position and living it. As I have already indicated, I think this distinction makes medical relativism untenable even if it is defensible.

Yet nor can medical relativism be dismissed. The fact that the dominance of Mainstream Medicine can be explained politically and socially needs to be addressed. That need is even more pressing in circumstances where the "hegemony" of "Western knowledge" is being actively called into question through calls for the "decolonization of knowledge."

9.2 WHAT IS DECOLONIZATION OF KNOWLEDGE?

Decolonization is originally a political concept, describing the process by which regions that were formerly part of the 19th-century

European empires become autonomous nation-states. In this usage its meaning is quite clear. However, it is also used in a different and largely metaphorical way, in phrases like "decolonization of knowledge" or "free, decolonized education." Over the last couple of years, these phrases have become a feature of the higher education landscape in contemporary South Africa, where I am writing. Many universities in the country now have committees, meetings, charters, and so forth devoted to decolonization. Yet South Africa is an autonomous nation, and, since 1994, a democracy. What do these calls mean?

As with political slogans in general, these uses of the term "decolonization" are not usually well defined. Calls to decolonize higher education refer both to the content of the curriculum and the manner in which it is taught, as well as the structures of the university, the staff complement, and the cost of higher education. The debate about access and financial barriers to higher education has global resonance, but the debate about decolonization is much more contextual. It means both increasing the access of African students to higher education, and reducing the influence of Western/Northern culture on what is taught, and perhaps on how it is taught, and increasing the influence of local—in this context, African—ideas. These meanings are in tension, in higher education and in medicine, and I will explore this tension in the next two sections.

Medicine is undoubtedly subject to cultural influences. Medical beliefs may be bound up with or even central to a culture. They may concern ancestors, vital forces, God, a scientific worldview: beliefs we will not easily part with, yet cannot readily defend, because they are so basic to the way we think. Medical practices, on the other hand, are ultimately aimed at the goal of cure, and this, I have suggested in the first part of the book, is broadly a shared feature of medicine. All medicine has cure as the goal, I argued, and moreover health is a near-universal concept. The experience of being sick, or of being desperately worried about one's sick child, is as close as we can get to a universal human experience, in my view, and I simply struggle to believe anyone who professes to think otherwise. Moreover, while there are some cases of cultural variation about what *counts* as healthy and what diseased, there is broad agreement. Given that medicine is culturally imbued and yet deals with universal human experiences, and has a universal goal, what are we to make of the idea of decolonizing it?

In the next two sections, I will consider and reject two ideas about what decolonizing medicine might be, corresponding to the two issues identified so far, access and knowledge. Relating to access, decolonization of medicine might be a call for greater *fair access* in the availability and practice of medicine. However, this idea about decolonization assumes that what amounts to good medicine is settled, and of course this is not settled in this context. This leads to the other possibility, which is that decolonization is *epistemic equality*. Medical knowledge claims from different traditions should be treated equally, none prioritized above the other. The problem with this is that it forces us into a naïve relativism that is really not plausible, even if more sophisticated relativism is plausible. If it is a fact tuberculosis cannot be cured without removing the tubercle bacilli from the sufferer, and if sacrificing a goat has no effect on the presence of the tubercle bacilli, then sacrificing a goat to appease the ancestors is just not going to help the patient with TB (even if there are ancestors and even if they are appeased), whatever that person or her community may believe. In the remainder of the chapter I will return to the notion of decolonization and remodel it to Cosmopolitan taste, so as to yield a more fruitful approach to disagreement about medical practices, even when there are tricky political, social, and cultural factors at work.

9.3 DECOLONIZATION AS FAIRNESS

One way in which medicine might need decolonizing is in its availability. Access to medicine remains divided along race, class, and gender lines. In South Africa, there is constitutional provision for universal access to healthcare, but the infrastructure to enable this provision is not yet in place. Public hospitals are overcrowded and under-resourced. The quality of medical care may be very high in many of these hospitals, but the process of getting to see a doctor can be horrific. The situation is worse in many other African countries. There are private health insurance schemes, and many employers (including mine) require employees to belong to one. If you do, you usually have much better access to care (depending on how much you pay). The ratio of White South Africans to Black South Africans in private hospitals is much higher than in public ones. (I am using "Black South African" is an umbrella category for the three other official South African races, namely African, Indian, and Colored—the latter being a specific South African racial classification, not to be confused with its usually offensive American usage.) In this sense, medical care in South Africa still transmits the mark of South Africa's colonial past, and more recently, apartheid. Blacks as a group still have, on average, much worse access to healthcare than Whites as a group, regardless of the lifting of legal barriers to access, and the overriding reason for this situation is the previous existence of a legal system that was exploitative along racial lines, which in turn is explained by the fact that South Africa was a colony.

I use South Africa, and Africa more generally, as an example because I live here, and because it is a part of the world where the colonial heritage still weighs heavy. However, the same effect is detectable even *in* the clinic. In American contexts, African Americans, women, and people from lower socioeconomic strata are less likely to receive pain relief from a prescribing physician than high-status white males. Daniel Goldberg makes interesting historical connections between our attitudes to pain and mechanical thinking in 19th-century medicine (Goldberg 2017). Technologies like X-rays, in particular, helped some pain sufferers, but not others. They helped those whose pain corresponded with something you could see on an X-ray, but not the rest. Those still had to rely on being able to convince a doctor that they were in pain. And your ability to do this relates to your race, gender, socioeconomic class, and many other things: your weight (fat people are not taken as seriously), your personality, and so forth. Either African Americans experience significantly less pain than white Americans, women less pain than men, poor people less pain than rich, and so forth, or there is an uneven level of prescription in relation to the same level of pain. The former is extremely implausible.

Pain is interesting because it is fundamentally subjective, and for this reason, its clinical management tells us something about patient credibility. If someone tells you they are in pain then you have no independent way to verify how much pain they are in. Even when we can do some sort of objective inspection, it does not tell us much: a cut can look bad but not hurt much, and a little scrape can be extremely painful despite not looking bad. An MRI scan can reveal a herniated disc, but it is possible to have disc degradation on an MRI scan and feel no pain, while others have pain but no herniated disc, and no other objectively detectable problem (literature on medically unexplained symptoms). Because of its subjectivity, pain displays clinical attitudes to its sufferers.

I assume that this broad empirical point is accepted: there is unequal access to healthcare, both in accessing the clinic and in accessing medical treatment after entering the clinic, and this unequal access follows the tediously familiar lines of race, gender, and socioeconomic class.

The obvious prescription in response to this social malaise
double efforts to universalize access. In South Africa, and many other
places (probably including America), more needs to be invested in
public hospitals, or else realistically affordable insurance schemes
need to be developed, or some other public initiative to solve this
problem needs to be undertaken. In America, and probably everywhere else too, clinicians need to be made aware of their statistical
tendencies to take certain people more seriously than others for no
good reason.

The difficulty with this suggestion in the current context is that
the suggestion itself is apt to yield results that have a colonial flavor
of their own. The downside of universal provision is that it supposes
that someone knows best. The people who do the providing also decide what to provide. We may be equalizing in one respect—equality
of access—but the underlying power relations remain along the same
lines as they were, and may be amplified. The state gets bigger, more
intrusive, and more involved in shaping the lives of individuals and
communities. Othering continues: "we" end up deciding what "they"
need and what "they" will get, where "we" are the social and political
elite, and "they" are the rest, who get our sympathy and our assistance,
but who "we" still do not see as equals, even if "we" will not say so.

This is a general problem, but it is at its most obvious when the
system of medicine favored by the elite is culturally and historically
quite different from the system of medicine that predated that elite.
In South Africa, the lack of Black access to Mainstream Medicine
during apartheid was a direct consequence of policy, and in consequence, even as South Africa was claiming credit for hosting the
first heart transplant, traditional African medicine remained the only
medicine available to many people. Today, this remains true in many
rural areas, in many parts of Africa. Everyday medical care will be
undertaken by "traditional healers"—an awkward term since "they

are neither particularly traditional nor healers as these terms are usually used" (Thornton 2009, 31).

The consequence is problematic if you start from the point of view that Mainstream Medicine is superior to traditional African medicine. But why should you start from that perspective? I have emphasized in the previous chapter the extent to which I believe that the testimony of people accepted as experts plays a role in persuading us to accept a medical practice or tradition. This is so, I have emphasized, even in relation to the kind of evidence favored by EBM; for after all, which of us has conducted an RCT, or would be able to tell a good one from a bad one merely by reading about it, if one lacked context—knowledge of personal or institutional reputations, knowledge of relevant literature, and so forth? Ultimately even a report of a trial requires trust in the honesty of the author, the competence of the experimenters, the fairness and rigor of the publication process, and so forth (trust that some of the critics of EBM suggest is misplaced, as we saw in Chapter 3). In short, who you believe determines what you accept, to a very large extent.

In this context, the assertion that Mainstream Medicine is superior to traditional African medicine *can* be, or become, an announcement of the assertor's predisposition (prior to evidence or reasoning) for accepting the pronouncements of one culture over another. This is particularly problematic if it becomes the basis for public policy, and for enforcing one's own ideas about which culture is better—which usually, but not necessarily, would be one's own—on people who do not share your preferences. This is another form of subjugation, to use that language; it is epistemic subjugation, and may persist *through* apparently equitable and public-spirited measures. Most obvious is the provision of an education, featuring certain things and not others—the history of Europe and not Africa, for the obvious example. And, runs the line of thought, another apparently equitable and public-spirited measure that might (surreptitiously) serve to subjugate one culture to another could be the widening of access to a certain kind of healthcare, probably accompanied by education programs that denigrate other traditions, and perhaps also threaten, or are perceived to threaten, ways of life that are regarded as traditional or culturally important. Medicine joins education and religion as tools in the colonizer's armory.

So runs this line of thought, at least, and it suggests another notion of decolonization, which we shall now consider.

9.4 DECOLONIZATION AS EPISTEMIC EQUALITY

Decolonization of medicine as fair and equal access to medicine is a laudable goal, but it may backfire if it is not accompanied by some deeper thinking about what counts as medicine in the first place. Otherwise, efforts to widen access might actually amplify and entrench existing power relations between cultures, racial groups, genders, and so forth. A few years ago, I met an American woman in Swaziland, working on a project whose goal was to circumcise as many Swazi men as possible. Circumcision, she told me, has been shown to reduce HIV infection rates in clinical trials: certainly not a complete protective effect, nor even a particularly large one, but a detectable one nonetheless. It is reasonable to be skeptical of this sort of evidence for this sort of intervention—that is, one that is going to be deployed in circumstances quite different to those of the trial, and in circumstances that an RCT probably does a poor job of informing us about (see section 5.5). In particular, compensatory changes in behavior such as a perceived need not to use a condom might cancel out any benefit, and the attempt to persuade people to undergo the procedure might well encourage them to overestimate its protective effect. An RCT, of course, would establish a protective effect *controlling for behavior*, but in practice this intervention could very well result in a change in behavior.

But even if we set aside questions of efficacy, the idea that one can simply fly off with this result in hand and set about persuading as many people as possible to submit to the procedure is, in my view, hard to defend. The power relations between the persuader and the persuaded, the relative poverty and lack of education of the persuaded party, the fact that circumcision is a culturally significant act that is practiced by some people in the region but not by all, the fact that a risk attends to any operation, notably risk of infection, the possible implication that African males' sexuality is uncontrollable and/or that African society is irredeemably misogynist and/or that Africans are incorrigibly promiscuous, so that the only feasible interventions are crude and physical ones, all trouble me.

Imagine what American women would make of a Swazi man who flew to Kansas to persuade them of the merits of polygamy, which is a traditional kinship pattern in many parts of Africa, but which in the United States is often regarded as embodying unacceptable patriarchy. Imagine that the men were typically students who spent at most a couple of years seeking to educate the locals of Kansas before heading back to Swaziland, most of them never to return. And imagine further that the entire exercise was funded by an organ of the United Nations. I would not approve of such an expenditure, and it would not change my mind to hear that, in RCTs conducted in Swaziland, polygamy had been shown to have some protective effect against (say) breast cancer (perhaps because it increases pregnancy rates—but this is purely an imaginary example). The cultural significance of polygamy is such that even if there *is* some such effect established in some trial, it is not right to simply try to export the practice to Kansas.

None of these motivations amount to a real chain of argument; they are suggestive only. But however we get there, one way to understand decolonization of medicine might be in terms of the *epistemic equality* of different medical traditions. On this view, the knowledge claims of different medical traditions must be taken equally seriously. To be even slightly plausible, this view must include the nature of evidential support in its purview. Both epidemiological evidence and reasoning from biological knowledge count as products of the same culture that produces Mainstream Medicine, so small wonder—the thinking goes—that they favor Mainstream Medicine. If we instead give attention to the kinds of evidence adduced by other traditions, we will likewise find little support for Mainstream Medicine. We are not evaluating particular claims against each other, but rather whole systems or worldviews, or—in the frustratingly inaccurate jargon of the contemporary debate—epistemologies.

The trouble with this way of approaching the matter is that it appears to lead us to relativism about medicine and medical knowledge; and, I shall now argue, wholesale medical relativism is implausible.

It is useful to distinguish wholesale medical relativism from its much more plausible cousin, namely partial medical relativism. We can accept some degree of relativism in medicine, as in many other matters. It is plausible, for example, that a given mood in a person who has suffered a bereavement is not a disease, but simply normal grief, while the same mood in someone who has suffered no major life event either is or is a symptom of a disease, namely depression. It is likewise plausible that the effectiveness of an intervention might be relative to the context because the goal of the intervention for a given disease might differ with context. So, for example, if I wake up with a rotten cold and it is a weekend or holiday, I might just stay in bed, while if I wake up with the same cold on a day when I have something important to do, I will use medication to control the pain and sneezing, and do my best to soldier on. These interventions achieve different ends, and which end I seek depends upon the context, and of course the unavailability of a miracle cure, which would presumably be my first choice in any context. Partial relativism about what works, what counts as working, what counts as disease, and so forth is plausible.

But partial relativism of this kind is generally eliminable by being clearer about the context. Low mood is not, itself, either a disease or not a disease; rather, low mood in bereavement is grief (not a disease), while low mood with no obvious cause is depression (a disease). Bed rest is probably better for healing a cold, but painkillers allow me to undertake my important task without distraction from a headache, perhaps at some expense to my health, which I am fundamentally healthy enough to afford. Ultimately, there is no deep relativism here, just initial vagueness, which, when clarified, resolves into a set of agreed facts. I have not shown that all partial medical relativism can be resolved in this way, but being unable to think up durable examples to the contrary, I feel justified in assuming that much of it can.

Wholesale relativism, on the other hand, is not so plausible. When Thabo Mbeki, former president of South Africa, asserted that AIDS was a social disease, he also denied that it was caused by a virus [ref]. He was surely right that AIDS is a social disease, in the sense that it clearly has social determinants: that is why it is so geographically localized to sub-Saharan African, and so much more prevalent in some subpopulations than others even within the region. But he was wrong about the latter. Even if he had great reasons for what he said, even if it was an honest and well-meaning mistake, he was still wrong. The fact is that AIDS is caused by a virus, and he said that it was not, and he was wrong about that.

It is also very probably a fact that Mbeki's stance cost a large number of South African lives. Had he taken a different view, effective public health interventions would probably have begun earlier, and this would probably have resulted in fewer HIV infections and thus in lives that were lost to AIDS not being lost. This is not as certain as the fact that HIV causes AIDS, but we should not confuse epistemology with metaphysics: we should not confuse the question about whether there *is* a fact with the question about *how certain* we can be about that fact. There is some fact about whether Mbeki's views made a difference, because there is a fact about whether the policies he could have enacted but did not were effective policies or not.

In the context of medicine, a wholesale and simplistic relativism, a view that anything goes, is not plausible. Medicine has practical consequences. These practical consequences might admit of different conceptualizations, but the actual variety of these conceptualizations probably far undershoots their theoretical possibility. Kwame Anthony Appiah emphasizes that we are all human, and that if we did not share a great deal, there would not even be the possibility of understanding other cultures. Donald Davidson makes a similar point in quite a different context. Whatever this common experience is, it seems likely that medicine deals with at least some of it: the pain of disease, the fear of caring for a sick child, the prospect of one's own death, are all things that people broadly understand and agree about, regardless of culture. And the effectiveness of medicine ultimately relates to its ability to help us with these big, shared, human concerns. Relativism about whether boiled chicken feet make for a nice dinner is one thing; relativism about whether they cure your child's life-threatening asthma is altogether another, regardless of how you conceptualize asthma, chicken-feet soup, and so forth.

This is my first reason to reject wholesale medical relativism: that medicine deals in matters that are of universal human concern, and as such, can be measured in relation to those universal concerns, even if we cut the cake differently along the way. A second reason is suggested by Appiah's critique of wholesale moral relativism. Appiah points out that a consistent moral relativism does not motivate any particular stance toward competing moral views. In particular, it does not motivate tolerance of views with which one disagrees. Appiah mentions a story to illustrate the point:

> One may be reminded of an old story from the colonial days of India. A British officer who was trying to stop a suttee was told by an Indian man, "It's our custom to burn a widow on her husband's funeral pyre." To which the officer replied, "And it's our custom to execute murderers."

(Appiah 2007, 24)

The point is that nothing in particular follows from the fact that all views are on an equal footing.

In the medical context, even if we accept the relativistic idea that all medical traditions are epistemically equal, it does not follow that one cannot pursue one's own tradition where the others contradict it. Accepting that traditions are epistemically equal does not in fact lead to tolerance of multiple traditions, since tolerance is not necessarily what these traditions espouse. Medical relativism would immediately put an end to much traditional African medicine, Chinese medicine, Indian medicine, Alternative therapy, alongside much Mainstream Medicine. The most powerful medical force would dominate, and it would not be wrong to do so.

If this is decolonization of medicine, then decolonizing will have the paradoxical consequence of immediately closing down all the differing traditions that we had thought we might find a way to reconcile. Given that this cannot be the principle, then, we are at liberty to practice our own medical traditions where we can. And if one finds oneself in a position of power, where one is able to prevail in a conflict between traditions, the fact that there are other traditions does not in itself give one any reason not to stick to one's own tradition and to give consideration to others, just as the fact that burning widows was a tradition in India did not give the British colonial officer any reason to deviate from his own tradition—which he was in a position to enforce—of executing those who performed such deeds.

The position as I see it, then, is as follows. Decolonization is naturally understood as epistemic equality, in the context of medicine. But epistemic equality in a context where different standards of evidence are in play means some form of relativism. This is uncontentious if it is partial, since partial medical relativism can usually be dissolved by a closer specification of the context. It is very contentious if it is wholesale, however. This is because medicine deals in fundamentals of human experience that are probably universal and these provide a measure of effectiveness about which it is very hard to be a wholesale relativist. Moreover, even if epistemic equality is the view, and wholesale medical relativism is adopted, the result is not decolonized medicine, because there is nothing in relativism or epistemic equality to enforce respect of other traditions, equal weighting, or anything of the sort, and indeed any universal principle to this effect would have the consequence of immediately putting a stop to any medical practice that disagreed with any other, which would be paradoxical.

And yet there is still the fundamental difficulty we started with: medicine clearly has strong cultural components, even if it also has strong non-cultural ones; and in particular, the evidence for medicine is, in practice, expert judgement, which is a kind of evidence that depends heavily upon one's cultural background for its persuasive force.

9.5 DECOLONIZING DECOLONIZATION

One option at this stage is to declare that medicine cannot be decolonized. But that seems premature. It is wrong that access to medicine is so unequal, even if attempting to encourage equal access can exacerbate epistemic inequality; and it is wrong for social and cultural power relations—artifacts of the waxing and waning of the fortunes of different regions, peoples, empires, ideas—to give rise to an inequality that affects epistemic matters. However, it is not easy to see how to get it right, beyond urging a pragmatic kind of balance between, on one side, tolerance, open-mindedness, and willingness to learn, and on the other, decisiveness, adherence to principles, and willingness to make a call when it is clear that lives are on the line.

A Cosmopolitan analysis would advocate four things, as we have seen in the previous two chapters: a presumption that there is a fact of the matter; a commitment to epistemic humility; a commitment to treat interlocutors as moral equals; and a prioritization of practice. Much that I have said in relation to Alternatives can be applied equally to other traditions where there is conflict. I will not repeat those arguments here. What I will do, however, is seek to deconstruct the notion of decolonization that led to this impasse, and remodel it in a way that makes it not only more coherent but also more respectful of the kinds of indigenous knowledge that it purports to rehabilitate.

Decolonization is sometimes presented, not as an attempt to resurrect the dispassionate search for knowledge, but as *a rejection of the idea of objectivity* which is seen as a sort of heritage of colonial thinking. In short, decolonization is sometimes equated to epistemic relativism. Another common term is *epistemic pluralism*, but, while philosophers distinguish between pluralism and relativism, it is not at all clear that there is any difference in the implications of these terms in this context. Both admit multiple "ways of knowing," in a way that makes it wrong to say—or deny—something like "It's just a fact that cholera is caused by *Vibrio cholerae*, and has nothing to do with ancestors."

Sometimes the idea is that notions like truth, fact, or what "works" are fundamentally Western and are imposed on other cultures. The

historical survey of Chapter 1 shows that this is unfounded for medicine, but in a political engagement such claims are typically not historically contextualized. At other times, the idea seems to be that
facts and truths are local, so what is discovered or expressed in one
time or place will necessarily be inapplicable in another.

This line of thinking is political; to be precise, it is part of a neo-
Marxist ideology, in which Marx, Foucault, and words like "post-
colony" and "hegemony" feature heavily. On this line of thought,
the attempt to critically evaluate the opinion of another person
or group looks like an exercise in power politics. It is a short step
from there to the idea that, in order to rid ourselves of the effects
of a colonial past, we must all desist from asserting our beliefs
over others' beliefs. There is African belief, and European belief,
and your belief, and mine—but none of us have the right to assert
that something is true, is a fact, or works, contrary to anyone else's
belief.

On this view, to decolonize knowledge is to understand this and
so to adopt a certain very broad kind of relativism. And logic suggests
that, on this view, decolonizing medicine would amount to rejecting
any universal claims of effectiveness, along with any universal claims
of theoretical correctness. Antibiotics are one way to treat tuberculosis, ritual prayer another; the tubercle bacilli cause TB so far as
I'm concerned, but for you, it could be ancestors. Unto each his own
medicine.

One very simple reason to reject relativistic decolonization is that
the kind of relativism I have described is associated with traditions
of thought that are characteristically European. Marxism and post-
modernism are about as European as baguettes and sauerkraut. Even
if some of their protagonists were born in Algeria, it was postwar
Europe that provided the soil for these ideas to grow. It would be a
different matter if we were being asked to accept the teachings of a
great African relativist. But we are not.

And this leads to a second reason to reject relativistic decolonization: that it is disrespectful to real disagreements. Where ideas conflict, that is typically a point of agreement between disagreeing parties (or would be, if they were all alive and talking). Far from being asked to accept the teachings of an African relativist, we are normally being asked to accept the teachings of African absolutists, alongside absolutists from Europe and other places; and so we are being asked to do an injustice to them all, since thinkers do not typically regard their thoughts as true only for them, or from their perspective.

A third reason to reject relativistic decolonization arises if one wants to avoid conflict by resolving disagreements. Appiah values conversation very highly as a way to resolve disagreements, and he rejects relativism in part because it doesn't motivate conversation. As he puts it:

> If we cannot learn from one another what is right to think and
> feel and do, then conversation between us will be pointless.
> Relativism of that sort isn't a way to encourage conversation; it's
> just a reason to fall silent.
>
> (Appiah 2007, 31)

This is a serious problem for relativistic decolonization, whose great attraction is that it appears to prevent us coming into conflict with each other, by preventing us from really disagreeing in the first place: instead, we simply accept that everyone is entitled to their views. But a moment's thought shows that there is no protection against possible conflict. I might believe I can do things you don't want me to do. I might even have views about your views.

But there is another way to think about decolonization, not as an invitation to relativism, but to *criticism*. No longer do we accept the contents of the canon: we realize that it has been influenced by who won what war, and with this in mind we ask whether its contents are really the best pieces of thought, should really be taught at the expense of all else, and so forth.

This sort of *critical decolonization* means accepting risk of error. It means considering whether the canon might be wrong: maybe Hume was a bad philosopher, maybe Shakespeare was a bad poet. But it also means asking whether ancestors really cause disease, or care about slaughtered goats. Indigenous knowledge systems might contain truths that Western science has not accessed. This is uncomfortable for Western science. But indigenous knowledge systems might also be wrong, either partially and understandably, or totally and extravagantly. This might be very uncomfortable for those knowledge systems. The goal of critical decolonization is to get to the truth, and nobody gets a free pass. Nor should they; after all, why should anyone, doctor or *sangoma*, student or professor, author or reader, president or farmer, be entitled to assert something without question or challenge, merely because it will upset them to be told that they are wrong?

I strongly suspect that the widespread appeal of decolonization slogans arises from equivocating between the relativistic and critical notions of decolonization. The term itself suggests the critical attitude: we should no longer accept the canon, but should open our eyes to the political influences that gave rise to it, adopt a critical attitude toward it, and cast the net much wider for new ideas. But then there is a subtle substitution of ideas. We should not only question the canon; we should *reject* it, because we can see that it is the child of unfair power relations. We must likewise disavow any further kinds of epistemic hegemony or violence, allowing people to express their beliefs, and refraining from oppressing them, even under the name of reason, logical consistency, empirical efficacy, or similar. These are just tools of epistemic oppression.

I strongly suspect that the appeal of calls for decolonization arises from the eminent attractiveness of the former idea, and the subtle substitution of the latter. Because epistemic relativism is defensible, it is very hard to resist the latter idea once the substitution has been made; and if one resists it, one can easily sound (or be made to sound) like one is also resisting the former idea, the celebration of critical attitude, which is tantamount to admitting that one is a dogmatist. Thus is philosophy pressed into the service of politics.

A final argument in favor of critical decolonization is that it enables us to resolve the tension between improving access to medicine and pursuing epistemic equality. We pursue epistemic equality by adopting the stance of epistemic humility, not the stance of epistemic relativism. Disagreements are resolved (starting with disagreements about practice, as emphasized in the previous chapter) and, with them, the tension between decolonization as fair access and as epistemic equality. This is an ideal, but at least it is accompanied by a recipe for achieving it; and at least it is a coherent and desirable ideal.

9.6 CONCLUSION

Decolonizing medicine could mean either improving access to medicine or removing the epistemic inequality between different traditions. These goals are in tension, because more access implies a decision about what medicine to make accessible. This tension cannot be resolved without critically assessing the idea of decolonization itself. That idea can be understood either relativistically or critically. There are good reasons not to adopt relativistic decolonization: it is a characteristically European idea; it fails to respect the reality of disagreements and the seriousness of beliefs; and it fails to encourage conversation, thus failing to avoid conflict. Critical decolonization, on the other hand, requires that knowledge claims be critically evaluated, regardless of their canonical status.

Critical decolonization is an implementation of the epistemic stance of Cosmopolitanism, which is epistemic humility: the willingness to change one's mind in a disagreement with people or evidence. If one adopts this attitude then it follows that, regardless of whether one regards a belief as canonical or not, one will subject it to critique in some circumstances.

REFERENCES

Ananth, M. 2008. *In Defense of an Evolutionary Concept of Health: Natures, Norms and Human Biology*. Aldershot, UK: Ashgate Publishing Ltd.

O Appiah, Kwame Anthony. 2007. *Cosmopolitanism: Ethics in a World of Strangers*. London: Penguin Random House.

Barnes, Barry, and David Bloor. 1982. "Relativism, Rationalism and the Sociology of Knowledge." In *Rationality and Relativism*, edited by Martin Hollis and Steven Lukes, 21–47. Cambridge, MA: MIT Press.

Beauchamp, Tom L., and James F Childress. 2013. *Principles of Biomedical Ethics*. 7th ed. New York: Oxford University Press.

Berlin, Brent, and Paul Kay. 1999. *Basic Color Terms: Their Universality and Evolution*. Center for the Study of Language and Information, Stanford University.

Bloor, David. 1982. "Durkheim and Mauss Revisited: Classification and the Sociology of Knowledge." *Studies in History and Philosophy of Science* 13 (4): 267–97.

Bloor, David. 1991. *Knowledge and Social Imagery*. 2nd ed. Chicago: University of Chicago Press.

Bloor, David. 2007. "Epistemic Grace: Anti-Relativism as Theology in Disguise." *Common Knowledge* 13: 250–80.

Bloor, David. 2008. "Relativism at 30,000 Feet." In *Knowledge as Social Order: Rethinking the Sociology of Barry Barnes*, edited by Massimo Mazzotti, 13–33. Aldershot, UK: Ashgate.

Boorse, Christopher. 1977. "Health as a Theoretical Concept." *Philosophy of Science* 44: 542–73.

Boorse, Christopher. 1997. "A Rebuttal on Health." In *What Is Disease?*, edited by James M. Humber and Robert F. Almeder, 1–134. Biomedical Ethics Reviews. Totowa, NJ: Humana Press Inc.

Boorse, Christopher. 2011. "Concepts of Health and Disease." In *Handbook of the Philosophy of Medicine*, edited by Fred Gifford, 16:13–64. Amsterdam: Elsevier.

Boorse, Christopher. 2014. "A Second Rebuttal on Health." *Journal of Medicine and Philosophy* 39 (6): 683–724.

Boorse, Christopher. 2016. "Goals of Medicine." In *Naturalism in the Philosophy of Health: Issues and Implications*, edited by Elodie Giroux, 145–77. Switzerland: Springer.

Broadbent, Alex. 2011a. "Defining Neglected Disease." *BioSocieties* 6 (1): 51–70.

Broadbent, Alex. 2011b. "Epidemiological Evidence in Proof of Specific Causation." *Legal Theory* 17: 237–78.

Broadbent, Alex. 2013. *Philosophy of Epidemiology*. New Directions in the Philosophy of Science. London and New York: Palgrave Macmillan.

Broadbent, Alex. 2015a. "Causation and Prediction in Epidemiology: A Guide to the 'Methodological Revolution.'" *Studies in History and Philosophy of Science Part C: Studies in History and Philosophy of Biological and Biomedical Sciences* 54: 72–80.

Broadbent, Alex. 2015b. "Epidemiological Evidence in Law: A Comment on Supreme Court Decision 2011Da22092, South Korea." *Epidemiology and Health* 37: e2015025.

Broadbent, Alex. 2015c. "Risk Relativism and Physical Law." *Journal of Epidemiology and Community Health* 69 (1): 92–94.

Broadbent, Alex. 2016. *Philosophy for Graduate Students: Metaphysics and Epistemology*. London and New York: Routledge.

Broadbent, Alex. 2018a. "Intellectualizing Medicine: A Reply to Commentaries on 'Prediction, Understanding and Medicine.'" *Journal of Medicine and Philosophy* 43 (3): 325–41.

Broadbent, Alex. 2018b. "Prediction, Understanding and Medicine." *Journal of Medicine and Philosophy* 43 (3): 289–305.

Broadbent, Alex. forthcoming. "Health as a Secondary Property." *British Journal for the Philosophy of Science*.

Broadbent, Alex, and S.-S. Hwang. 2016. "Tobacco and Epidemiology in Korea: Old Tricks, New Answers?" *Journal of Epidemiology and Community Health* 70 (6): 527–28.

Broadbent, Alex, Jan P. Vandenbroucke, and Neil Pearce. 2016. "Response: Formalism or Pluralism? A Reply to Commentaries on 'Causality and Causal Inference in Epidemiology.'" *International Journal of Epidemiology* 45 (6): 1841–51.

Bynum, William. 2008. *The History of Medicine: A Very Short Introduction* [e-book]. Oxford: Oxford University Press.

Caplan, A. 1992. "If Gene Theory Is the Cure, What Is the Disease?" In *Gene mapping: Using law and ethics as guides,* edited by G. Annas and S. Elias, 128–41. New York: Oxford University Press.

Carnap, Rudolf. 1947. "On the Application of Inductive Logic." *Philosophy and phenomenological research* 8 (1): 133–48.

Cartwright, Nancy. 2007a. "Are RCTs the Gold Standard ?" *Biosocieties* 2: 11–20.

Cartwright, Nancy. 2007b. *Hunting Causes and Using Them: Approaches in Philosophy and Economics.* New York: Cambridge University Press.

Cartwright, Nancy. 2012. "Will This Policy Work for You? Predicting Effectiveness Better: How Philosophy Helps." *Philosophy of Science* 79 (5): 973–89.

Cartwright, Nancy. 2011. "Predicting What Will Happen When We Act. What Counts for Warrant?" *Preventive Medicine* 53 (4–5): 221–24.

Cartwright, Nancy, and Jeremy Hardie. 2012. *Evidence Based Policy: A Practical Guide to Doing It Better.* New York: Oxford University Press.

Cooper, Rachel. 2002. "Disease." *Studies in History and Philosophy of Biological and Biomedical Sciences* 33: 263–82.

Duhem, Pierre. 1914. *The Aim and Structure of Physical Theory.* New York: Atheneum.

Dummett, Michael. 1978. "Bringing About the Past." In *Truth and Other Enigmas,* edited by Michael Dummet, 333–50. Duckworth.

Eardley S, Bishop FL, Prescott P, Cardini F, Brinkhaus B, Santos-Rey K, Vas J, von Ammon K, Hegyi G, Dragan S, Uehleke B, Fonnebo V, Lewith G: A systematic literature review of complementary and alternative medicine prevalence in EU. Forsch Komplementmed. 2012, 19 (Suppl 2): 18–28.

Eisenberg, David M., Roger B. Davis, Susan L. Ettner, Scott Appel, Sonja Wilkey, Maria Van Rompay, and Ronald C. Kessler. 1998. "Trends in Alternative Medicine Use in the United States, 1990–1997: Results of a Follow-up National Survey." *Journal of the American Medical Association* 280 (18): 1569–75.

Ereshefsky, M. 2009. *Studies in History and Philosophy of Biological and Biomedical Sciences* 40: 221–27.

Ernst, E. 2000. "Prevalence of Use of Complementary/Alternative Medicine : A Systematic Review." *Bulletin of the World Health Organization* 78 (2): 252–57.

Evans, Imogen, Hazel Thornton, Iain Chalmers, and Paul Glasziou. 2011. *Testing Treatments: Better Research for Better Healthcare.* 2nd ed. London: Pinter & Martin.

Forester, John. 1996. "If P, Then What? Thinking in Cases." *History of the Human Sciences* 9 (3): 1–25.

Frank, Robert. 2002. "Integrating Homeopathy and Biomedicine : Medical Practice and Knowledge Production among German Homeopathic Physicians." *Sociology of Health and Illness* 24 (6): 796–819.

Frankfurt, Harry G. 2005. *On Bullshit.* Princeton, NJ: Princeton University Press.

Gabbay, Dov M., Paul Thagard, and John Woods. 2011. *Handbook of Philosophy of Medicine*, edited by Fred Gifford. Vol. 16. Handbook of Philosophy of Science. Amsterdam: Elsevier.

Goldacre, Ben. 2011. "Foreword." In *Testing Treatments: Better Research for Better Healthcare*, edited by Imogen Evans, Hazel Thornton, Iain Chalmers, and Paul Glasziou, 2nd ed. London: Pinter & Martin.

Goldberg, Daniel S. 2017. "Pain, Objectivity and History: Understanding Pain Stigma." *Medical Humanities*, 43 (4): 238–43.

Goldenberg, Maya J. 2016. "Public Misunderstanding of Science? Reframing the Problem of Vaccine Hesitancy." *Perspectives on Science* 24 (5): 552–81.

Goldman, Alvin I. 2007. "Experts : Which Ones Should You Trust ?" *Philosophy and Phenomenological Research* 63 (1): 85–110.

Gutting, Gary. 2014. "Michel Foucault." In *The Stanford Encyclopedia of Philosophy*, edited by Edward N. Zalta, Winter 2014. Metaphysics Research Lab, Stanford University. https://plato.stanford.edu/archives/win2014/entries/foucault/.

Hansen, Kirsten, and Kappel Klemmens. 2016. "Complementary/Alternative Medicine and the Evidence Requirement." In *The Routledge Companion to Philosophy of Medicine*, edited by Miriam Solomon, Jeremy R. Simon, and Harold Kincaid, 257–68. New York and London: Routledge.

Hare, R. M. 1975. "Abortion and the Golden Rule." *Philosophy and Public Affairs* 4: 201–22.

Harris, Chadwin. 2018. "The Continuing Allure of Cure." *Journal of Medicine and Philosophy* 43 (3): 313–24.

Hausman, Daniel. 1998. *Causal Asymmetries*. Cambridge, UK: Cambridge University Press.

Hemilä, Harri. 2009. "Vitamins and Minerals." In *Common Cold*, edited by Ronald Eccles, Olaf Weber, 275–307. Switzerland: Birkhauser Verlag.

Howick, Jeremy. 2017. "Justification of Evidence-Based Medicine Epistemology." In *The Bloomsbury Companion to Contemporary Philosophy of Medicine*, edited by James A. Marcum, 113–46. London and New York: Bloomsbury.

Kingma, Elselijn. 2007. "What Is It to Be Healthy?" *Analysis* 67 (2): 128–33.

Kingma, Elselijn. 2010. "Paracetamol, Poison, and Polio: Why Boorse's Account of Function Fails to Distinguish Health and Disease." *British Journal for the Philosophy of Science* 61: 241–64.

Kingma, Elselijn. 2014. "Naturalism About Health and Disease: Adding Nuance for Progress." *Journal of Medicine and Philosophy* 39 (6): 590–608.

Kling, J. 1998. "From Hypertension to Angina to Viagra." *Modern Drug Discovery* 1: 31–38.

Kusch, M. 2006. *Knowledge by Agreement*. Oxford: Oxford University Press.

Laudan, Larry. 1981. "A Confutation of Convergent Realism." *Philosophy of Science* 48 (1): 19–49.

Leibovici, Leonard. 2001. "Effects of Remote, Retroactive Intercessory Prayer on Outcomes in Patients with Bloodstream Infection: Randomised Controlled Trial." *British Medical Journal* 323: 1450–51.

Lewis, David. 1973a. *Counterfactuals*. Cambridge, MA: Harvard University Press.

Lewis, David. 1973b. "Counterfactuals and Comparative Possibility." *Journal of Philosophical Logic* 2: 418—446.

Lewis, David. 1983. "New Work for a Theory of Universals." *Australasian Journal of Philosophy* 61 (4): 343–77.

Lexchin, J., L. A. Bero, and B. Djulbegovic. 2003. "Pharmaceutical Industry Sponsorship and Research Outcome and Quality: Systematic Review." *British Medical Journal* 326: 1167–70.

Lipton, Peter. 2004. *Inference to the Best Explanation*. 2nd ed. London and New York: Routledge.

Lloyd, Geoffrey Ernest Richard. 1983. *Hippocratic Writings*. London: Penguin.

Locke, John. 1706. *An Essay Concerning Human Understanding*. 27th ed. London: Tegg and Son.

MacMahon, B., and T. F. Pugh. 1960. *Epidemiologic Methods*. Boston: Little, Brown.

Marmot, Michael. 2006. "Health in an Unequal World: Social Circumstances, Biology, and Disease." *Clinical Medicine* 6 (6): 559–72.

Marx, Karl, Friedrich Engels, and C. J. (Christopher John) Arthur. 1970. *The German Ideology*. London: Lawrence & Wishart.

Menzies, Peter, and Huw Price. 1993. "Causation as a Secondary Quality." *British Journal for the Philosophy of Science* 44 (2): 187—203.

Metz, Thaddeus. 2018. "Medicine Without Cure: A Cluster Analysis of the Nature of Medicine." *Journal of Medicine and Philosophy* 43 (3): 306–12.

Metz, Thaddeus, and Gaie, Joseph B. R. 2010. "The African Ethic of Ubuntu/ Botho: Implications for Research on Morality." *Journal of Moral Education* 39: 273–90.

Miles, Stephen H. 2005. *The Hippocratic Oath and the Ethics of Medicine*. New York: Oxford University Press.

Morabia, Alfredo. 2004. *History of Epidemiologic Methods and Concepts*. Basel: Birkhauser Verlag.

Murphy, Dominic. 2015. "Concepts of Disease and Health." In *The Stanford Encyclopedia of Philosophy*, edited by Edward N. Zalta. http://plato.stanford.edu/archives/spr2015/entries/health-disease/.

Neander, Karen. 2006. "The Teleological Notion of 'Function.'" *Australasian Journal of Philosophy* 69 (4): 454–68.

Ngubane, Harriet. 1977. *Body and Mind in Zulu Medicine: An Ethnography of Health and Disease in Nyuswa-Zulu Thought and Practice*. London, New York, and San Francisco: Academic Press.

Nozick, Robert. 1981. *Philosophical Explanations*. Cambridge, MA: Harvard University Press.

Okasha, Samir. 2008. *Evolution and the Levels of Selection*. Wotton-under-Edge, UK: Clarendon Press.

Pauling, Linus. 1970. "Evolution and the Need for Ascorbic Acid." *Proceedings of the National Academy of Sciences of the United States of America* 67(4) (December): 1643–8.

Peacock, Janet L., and Philop J. Peacock. 2011. *Oxford Handbook of Medical Statistics*. Oxford: Oxford University Press.

Porter, Roy. 1997. *The Greatest Benefit to Mankind: A Medical History of Humanity from Antiquity to the Present*. London: Harper Collins Publishers.

Porter, Roy. 2002. *Blood and Guts: A Short History of Medicine*. London: Penguin.

Price, Huw. 1996. *Time's Arrow and Archimedes' Point*. Oxford: Oxford University Press.

Price, Huw, and Richard Corry, eds. 2007. *Russell's Republic Revisited: Causation, Physics, and the Constitution of Reality*. Oxford: Oxford University Press.

Quine, W. V. O. 1953. "Two Dogmas of Empiricism." In *From a Logical point of View*, 20–43. Cambridge, MA: Harvard University Press.

Rockett, R. H. 1999. "Population and Healh: An Introduction to Epidemiology." *Population Bulletin* 54: 1–44.

Sackett, David L. 1997. *Evidence-Based Medicine: How to Practice and Teach EBM*. Edinburgh: Churchill Livingstone.

Sackett, David L., and William M. C. Rosenberg. 1995. "On the Need for Evidence-Based Medicine." *Journal of Public Health Medicine* 17 (3): 330–34.

Schroeder, S. Andrew. 2013. "Rethinking Health: Healthy, or Healthier Than?" *British Journal for the Philosophy of Science* 64 (1): 131–59.

Singh, Simon, and Edzard Ernst. 2008. *Trick or Treatment? Alternative Medicine on Trial*. London: Bantam Press.

Sisti, D., and A. L. Caplan. 2017. "The Concept of Disease." In *The Routledge companion to philosophy of medicine*, edited by M. Solomon, J. R. Simon, and H. Kincaid, 5–15. New York/London: Routledge.

Smart, Benjamin. 2016. *Concepts and Causes in the Philosophy of Disease*. Palgrave Pivot. Basingstoke: Palgrave Macmillan.

Smith, G. C., and J. P. Pell. 2003. "Parachute Use to Prevent Death and Major Trauma Related to Gravitational Challenge: Systematic Review of Randomised Controlled Trials." *British Medical Journal* 327 (7429): 1459.

Stalnaker, Robert. 1981. "A Defense of Conditional Excluded Middle." In *Ifs*, edited by W. L. Harper, R. Stalnaker, and G. Pearce, 87–104. Dordrecht, Holland: D. Reidel Publishing Company.

Stegenga, Jacob. 2018. *Medical Nihilism*. Oxford: Oxford University Press.

Stempsey, William E. 2000. *Disease and Diagnosis: Value-Dependent Realism*. Philosophy and Medicine. Dordrecht: Springer.

Tetlock, Philip. 2005. *Expert Political Judgement—How Good Is It? How Can We Know?* Princeton, NJ, and Oxford: Princeton University Press.

Thornton, Robert. 2009. "The Transmission of Knowledge in South African Traditional Healing." *Africa: Journal of the International Africa Institute* 79 (1): 17–34.

Vandenbroucke, Jan P. 2008. "Observational Research, Randomised Trials, and Two Views of Medical Science." *PLoS Medicine* 5 (3): e67.

Vandenbroucke, Jan P. 2016. Personal communication.

VanderWeele, Tyler J. 2015. *Explanation in Causal Inference: Methods for Mediation and Interaction*. New York: Oxford University Press.

Vickers, Andrew J. 2000. "The Journal of Alternative and Complementary Medicine. pdf." *Journal of Alternative and Complementary Medicine* 6 (1): 49–56.

Wakefield, J. (1992). The Concept of Mental Disorder: On the Boundary Between Biological Facts and Social Values. *American Psychologist* 47: 373–88.

Walach, Harald. 2003. "Reinventing the Wheel Will Not Make It Rounder: Controlled Trials of Homeopathy Reconsidered." *Journal of Alternative and Complementary Medicine* 9 (1): 7–13.

Warnock, Mary. 1984. *A Question of Life: The Warnock Report on Human Fertilisation and Embryology*. London: Her Majesty's Stationery Office.

Wootton, D. 2006. *Bad Medicine: Doctors Doing Harm Since Hippocrates* [e-book]. New York: Oxford University Press.

Worrall, John. 1989. "Structural Realism: The Best of Both Worlds?" *Dialectica* 43: 99–124.

INDEX